DIFFERENTIAL EQUATIONS AND MATHEMATICAL MODELLING

DIFFERENTIAL EQUATIONS AND MATHEMATICAL MODELLING

A.M. BLOKHIN

Nova Science Publishers, Inc.
New York

Senior Editors: Susan Boriotti and Donna Dennis
Coordinating Editor: Tatiana Shohov
Office Manager: Annette Hellinger
Graphics: Wanda Serrano
Editorial Production: Jennifer Vogt, Matthew Kozlowski, Jonathan Rose and Maya Columbus
Circulation: Ave Maria Gonzalez, Indah Becker, Raymond Davis, Vladimir Klestov and Jonathan Roque
Communications and Acquisitions: Serge P. Shohov
Marketing: Cathy DeGregory

Library of Congress Cataloging-in-Publication Data
Differential Equations / A.M. Blokhin (editor)
p. cm.
Includes index.
ISBN 1-59033-085-4
1. Differential equations. I. Blokhin, A.M. (Aleksandir Mikhailovich)
QA371.D4432 2001
515'.35--dc21 2001054693

227 Main Street, Suite 100
Huntington, New York 11743
Tele. 631-424-NOVA (6682) Fax 631-425-5933
E Mail: Novascience@earthlink.net
www.novapublishers.com

Printed in the United States of America

Contents

Chapter 1

On a certain class of nonlinear forward backward parabolic equations[1]

R.R. Akhmerov[2]

Abstract. The review of results concerning the nonlinear forward backward parabolic equations is given.

Introduction

The equation

$$u_t + uu_x = [\omega(u_x)]_x, \tag{1}$$

where $u: [0,1] \times [0,T] \longrightarrow \mathbb{R}$ is an unknown function, $u_x = \dfrac{\partial u}{\partial x}$, $u_t = \dfrac{\partial u}{\partial t}$, and $\omega: \mathbb{R} \longrightarrow \mathbb{R}$ is a given sufficiently smooth function, such that ω' changes the sign, was brought into mathematical practice by Yanenko [51] in 1973. He hoped that this equation will allow to simulate complex turbulente flows of viscous incompressible fluids. Later, the similar equation as mathematical object was introduced (it seems to be independent) by Höllig & Nohel [118], and also by Barenblatt (the reference to seminar discussion see in Dzhumagazieva [13]).

Depending on the $\omega'(u_x)$ sign, the equation (1) on the particular solution can be forward parabolic equation, or backward parabolic one.

The processes with a negative viscosity are observed for a long time and in the most diverse areas: from chemotaxis of microorganisms up to turbulent flames, from metastable chemical reactions up to problems of a landscape formation, from population genetics up to meteorology, from metallurgy up to numerical methods for continuous media problems. We especially note among these applications acoustic waves in nonequilibrium molecular gases, shock waves in boiling fluids, processes of circulation in atmosphere, thermodynamics of materials with fading memory, phase transitions with supercooling and superheating, dynamics of phase separation of alloys, processes of polymer's segregation, first

[1]This work was partially supported by Programme "Fundamental Research in Mathematics" Grant of Higher Education State Committee of Russian Federation.

[2]Institute of Computational Technologies, Siberian Branch of Russian Academy of Sciences, Lavrentjeva ave. 6, Novosibirsk 90, 630090, RUSSIA; e-mail: rust@ictnw.scnet.nsk.su, akhmerov@ict.nsk.su

differential approximation of Particles-in-Cells method for gas dynamics equations, hydrodynamics of fluids with complex reology and other. In the first paragraph we describe some of them. Wide applications cause natural growth of interest to the equation (1) as to purely mathematical object.

The theory of nonlinear forward backward parabolic equations (FBPEs) is rather far from completion. In the given paper we suggest the survey of results, relating to the equation (1) and its analogies. Because of the volume limitations we have not an opportunity to give the complete proofs, but always we strive to describe main ideas. The results received till 1982 adequately covered in the third chapter of the monograph Lar'kin, Novikov & Yanenko [25], and they are described only concisely, for preservation of a common outline of the paper. Various results devoted to the Cahn-Hilliard equation theory we describe only in part, so far as they concern to FBPE. Reasonably complete survey of these results can be found in Elliott [99]. We shall emphasize that in the review we restrict our interests in FBPEs. Their regularizations we consider just as regularizations, instead of independent object of investigation.

The structure of the paper is as follows. In the first Section we give applications giving birth to FBPEs. In the second one we describe preliminary results, in particular, we discuss *a priori* estimates of solutions. Section 3 contains detailed discussion of some selfsimilar problems (namely, stationary and traveling waves one). For these problems the most advanced results are obtained (certainly, because of their relative simplicity). A structure of solutions sets of these problems is described, compactness and closedness properties of these sets in the various functional spaces are discussed. The existence theorems for Dirichlet and Neumann boundary value problems (BVPs) for one-dimensional FBPEs are given in the fourth Section. We discuss there also the causes of ill-posedness of classical BVPs, conditions for their well-posedness (when the specified problems, in essence, appear to be parabolic one). In particular, an opportunity of passing to the limit in selfsimilar BVPs regularized by a small dispersion equation is studied. In the fifth Section various regularizations *per se* for FBPEs are described. Cahn-Hilliard equation have received particularly much consideration. And, finally, Section 6 is devoted to discussion of various passages to the limit with respect to a small parameter in regularizing terms.

References in the text, as a rule, mean a source of describing results, and not always reflect a priority.

Alongside with the equation (1) we shall consider also similar equation without convective term

$$u_t = [\omega(u_x)]_x. \tag{2}$$

For definiteness, we shall call the first equation *forward backward convection-diffusion equation*, and the second one *forward backward heat equation*. Besides, we shall consider the equation

$$v_t = [\omega(v)]_{xx}, \tag{3}$$

derived from (2) by differentiation in x and change of variable $u_x = v$. The equation (3) will be referred as *degenerate Cahn-Hilliard equation* (origin of this term will be explained below).

In the survey we continually attempt to consider extreme simple situation, when the problems keep a sense. As a rule, we restrict the exposition by one-dimensional equations, only mentioning the multi-dimensional generalizations. The formulations frequently weakened or are cited with the partial list of conditions, so the survey can be used as a reference paper only with care.

The author expresses his gratitude to Vasilios Alexiades, Donald S. Cohen, Charles M. Elliott, Michael Hazewinkel, Alan V. Lair, M.M. Lavrentjev, Jr., Luciano Modica, James D. Murray, Robert L. Pego, Marshall Slemrod, Peter Sternberg, Gilbert Strang and Hans Weinberger, for their kind agreement to send reprints at the author's request, and also G.G. Chernykh, N.N. Fedorova, V.N. Grebenev and P.I. Plotnikov for the useful information.

1. Forward backward parabolic equations in applications

In this Section we describe some natural science problems, causing to FBPEs.

1.1. Non-equilibrium phase transitions with supercooling and superheating

In the rectangular $Q = [0,1] \times [0,T]$ we consider nonlinear diffusion equation

$$e_t = \vartheta_{xx}, \tag{4}$$

where

$$\vartheta(x,t) = \omega[e(x,t)],$$

and $\omega\colon \mathbb{R} \longrightarrow \mathbb{R}$. If we treat e in the equation (4) as enthalpy and ϑ as temperature, then the function ω represents a nonlinear constitutive function. It is well known that if ω is strictly monotone outside of an interval (e_s, e_l) and has a constant value ϑ_p on this interval, the equation (4) describes heat transfer process in the presence of phase transitions (ϑ_p corresponds to the temperature of phase transition). In a number of situations such a process is accompanied by supercooling and superheating, i.e., by existence of areas of solid phase with temperature exceeding ϑ_p, and areas of liquid phase with temperature smaller then ϑ_p. In opinion of a number of investigators (see, f.e., Lacey & Shillor [122]) such (thermodynamically metastable) processes can be described by the equation (4) with non-monotone function ω (see, f.e., Elliott [92]). By the change of variable $e = u_x$ and integration in x it is obviously results to (2).

1.2. Structural phase separation and Cahn-Hilliard equation

A more detailed discussion of processes described in the previous subsection results to so called Cahn-Hilliard equation (CHE). Let in "one-dimensional" pipe of a unit length be a two-component solution (for definiteness, we denote these components by A and B) at temperature T, and critical temperature T_{cr} corresponds to thermodynamic equilibrium of the solution when two phases co-exist, one of which is enriched with A, and the second one with B. At temperature below critical a phase separation of the solution occurs with formation of "grains" enriched by the components A and B. For description of dynamics of two-phase mixtures CAHN & HILLIARD [83] has suggested in 1958 the equation

$$c_t + [J(c)]_x = 0, \tag{5}$$

in which c is a concentration of a component, say A, and $J(c) = [\sigma(c)]_x$ is a mass flux determined through the chemical potential σ; this potential, in turn, is defined as a gradient of Ginzburg-Landau generalized free energy

$$\mathcal{F}_\varepsilon(c) = \int_0^1 \left[W(c,T) + \frac{\varepsilon}{2}(c_x)^2\right] dx$$

(for the first time such an energy functional has been introduced by VAN DER WAALS [149]). The first term under the integral sign is a potential of Ginzburg-Landau free energy, and the second one represents a phenomenological description of energy of phase surface interaction induced by space concentration nonhomogeneity. Parameter ε characterizes an interaction length between components. If we denote $\frac{\partial}{\partial c}W(c,T)$ by ω, then (5) can be rewritten in the form

$$c_t + \varepsilon c_{xxxx} = [\omega(c)]_{xx}. \tag{6}$$

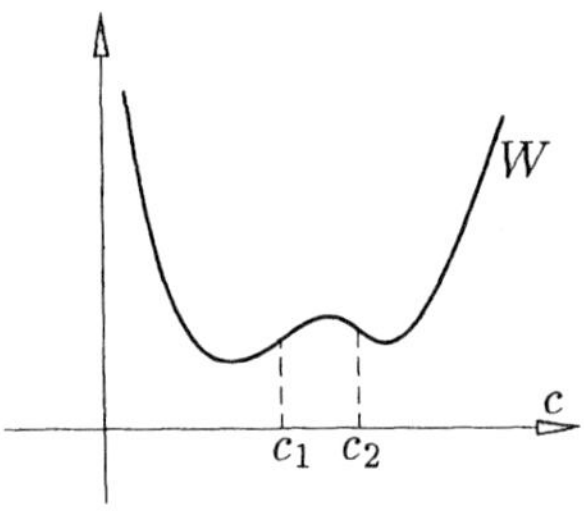

Fig. 1

In a number of solutions, at temperatures below critical one, the potential W is peculiarly shaped like "double-well" function (see Fig. 1). So, $\omega'(c) < 0$ on some interval (c_1, c_2) (called *spinodal interval*) and $\omega'(c) > 0$ on $(0,\infty)\backslash(c_1,c_2)$ (cf. with well-known Van der Waals potential). If we will not take into account the phase surface energy interaction ($\varepsilon = 0$), then we receive (after the change of variables $c = u_x$ and integration in x) the equation (2). An analogous transformation of CHE results to regularized FBPE

$$u_t + \varepsilon u_{xxxx} = [\omega(u_x)]_x.$$

The physical base of spinodal decomposition can be found, f.e., in SKRIPOV & SKRIPOV [45] and GUNTON & DROZ [109].

Multidimensional analogue of the equation (6) is the following equation

$$c_t + \varepsilon\Delta^2 c = \Delta[\omega(c)]$$

($c\colon \mathcal{D} \times [0,T] \to \mathbb{R}$, $\mathcal{D} \subset \mathbb{R}^m$).

The CHE represents a gradient dynamics equation for energy functional $\mathcal{F}_\varepsilon$ by suitable choice of a phase space. Namely, if $H_0^{-1}(\mathcal{D})$ is the subspace of the

Sobolev space $H^{-1}(\mathcal{D})$, selected by preservation of an average concentration condition

$$\int\limits_{\mathcal{D}} c(x)\, dx = m, \tag{7}$$

it follows that (see FIFE [120]) the gradient dynamics equation for $\mathcal{F}_\varepsilon$ on $H_0^{-1}(\mathcal{D})$ is CHE. The same equation on the subspace $L_{20}(\mathcal{D})$ of the space $L_2(\mathcal{D})$, selected by (7), represents the non-local Allen-Cahn model (see ALLEN & CAHN [66])

$$c_t = \varepsilon\Delta c - \omega(c) + \frac{1}{\mathrm{mes}\mathcal{D}} \int\limits_{\mathcal{D}} \omega[c(x)]\, dx.$$

The CHE is a special case of so called phase-field equations

$$a\vartheta_t + lc_t = k\Delta\vartheta,$$

$$\alpha c_t = \lambda\Delta c - \nu\omega(c) + l\vartheta.$$

These equations were suggested by CAGINALP [81,82], FIX [103] e.a., for modeling of a solidification processes in binary alloys. In these equations ϑ is a temperature, c is a phase field, a, l, k, α, λ and ν are positive constants. The first equation represents a heat equation, and the second one is the Ginzburg-Landau equation.

Obviously, if $a = 0$, $\alpha = 0$ and $l = 1$, then the phase field equations are turned to CHE

$$c_t = k\nu\Delta\omega(c) - k\lambda\Delta^2 c \ (= \Delta(k\nu\omega(c) - k\lambda\Delta c).$$

The phase field equations in the last years receive active attention. Here we restrict ourself only to mention just a few of works, as far as the mutual relations between phase field equations and FBPEs are far from understanding. Basically we present work containing extensive bibliography: ALIKAKOS & BATES [60], BATES & ZHENG [77], CHEN [86], DANILOV, OMEL'YANOV & RADKEVICH [12], GURTIN [110], PLOTNIKOV & STAROVOITOV [38], OMEL'YANOV, RADKEVICH & DANILOV [35], RADKEVICH [42], SONER [144].

In spite of the fact that the CHE is one of the most popular and physically consistent regularization of FBPE, this review can not be considered, in no account, as a review on this equation (in my estimation, bibliography concerning the theory of CHE consists of more than 200 items).

Cahn-Morall system of differential equations (see MORRAL & CAHN [127]) generalized CHE for multicomponent alloys. It was investigated in EYRE [101], GRANT [106], and ELLIOTT & LUCKHAUS [96], e.a. A generalization of the CHE for non-isotermic case is deduced and investigated in PENROSE & FIFE [137], ALT & PAWLOW [67] (see also ALT & PAWLOW [68], SHEN & ZHENG [141]).

1.3. Continuous version of Thom's river basin model

In THOM [148] the following model describing formation of strongly indented ravines' structures has been suggested. Down by sandy slope of a hill rain flow forms a system of flowing together and separating brooklets. Because of a water erosion this system changes structure in time. Let us consider a section transversal to the hill's gradient. Let it be parametrized by the length, and $s_n(t)$ be a position of the nth watershed between neighboring brooklets. We assume that the erosion power of a flow is proportional to its basin width with a factor k dependent on this width. Obviously, the watershed positions satisfy to the system of ordinary differential equations

$$s_n' = -k(s_{n+1} - s_n) \cdot (s_{n+1} - s_n) + k(s_n - s_{n-1}) \cdot (s_n - s_{n-1}).$$

Properly, this system is Thom's river basin model. In HAZEWINKEL, KAASHOEK & LEYNSE [114] has been shown that a continuous analogue of this model is the CHE

$$u_t + \varepsilon u_{xxxx} = [\omega(u)]_{xx},$$

where ω is a cubic polynomial such that ω' changes the sign.

1.4. A model of the ocean thermocline

A vertical heat transfer in an upper turbulente layer of the ocean can be described by the following equation (see, f.e., MONIN & YAGLOM [31])

$$T_t = [(K + kT_z)]_z; \tag{8}$$

where $T = T(z,t)$ is a temperature via depth z and time t, K and k are turbulent and molecular diffusion coefficients. In a series of models it is supposed that the turbulent temperature diffusion coefficient depends on the Richardson number (in other words, on gradient of temperature) in a nonlinear manner:

$$K = d(T_z).$$

In the simplest models $k = k_0 = \text{const}$. Some investigators consider as d the following approximation of experimental data:

$$d(p) = \frac{K_0}{1 + \alpha p^2}, \text{ or } d(p) = \frac{K_0}{(1 + \sigma p)^n}$$

($n = 2$ or $n = 3/2$; α and σ are positive constants). As far as $K_0 \gg k_0$, it is easy to show that the derivative of $\omega(p) = [d(p) + k_0]p$ changes the sign, and hence the equation (8) is a FPBE.

KRASOVSKII [23] explains (by physical reasoning) the occurrence of a layer structure in an ocean thermocline by alternating of ω' sign (the vertical profiles of temperature in these structures consist of interleaved regions with large and small gradients).

In BARENBLATT, BERTSCH, DAL PASSO & UGHI [71] and BARENBLATT, BERTSCH, DAL PASSO, PROSTOKISHIN & UGHI [70] the following regularized model of heat and mass transfer in steady stratified turbulent shear flows

$$u_t = [\omega(u_x)]_x + \varepsilon[\psi(u_x)]_{xt}$$

is discussed (this equation can be served as a regularization of the equation (8)). Here ω is a positive smooth function equal to zero at zero and infinity, and ψ is a strictly monotone smooth function, such that $\omega'(p) = O[\psi'(p)]$. The equation is derived under conjecture on the presence of a small non-zero relaxation time ε of turbulence field to a given gradient of temperature.

1.5. The first differential approximation of Particles-in-Cell method for gas dynamic equations

Particles-in-Cell method for wave equation was investigated in DALY [89]. The first differential approximation (see SHOKIN & YANENKO [49]) for this method represented in a form of explicit two-point central difference scheme for a modified differential equation can be rewritten as the system of equations

$$u_t + \frac{\tau}{2}u_{tt} = -pv_x + \left[\left(Ch|u| - \frac{1}{2}p^2\tau\right)u_x\right]_x, \tag{9}$$

$$v_t = -p^2 u_x,$$

where τ and h are the time and space steps, C and p are constants. Namely the alternation of the derivative of the function $u \mapsto Ch|u| - \frac{1}{2}p^2\tau$ sign DALY [89] explains appearance of oscillations and instability of the numerical method. The equation (9) in some sense can be considered as a FBPE.

1.6. Viscoelastic non-Newtonian fluids

In HUNTER & SLEMROD [116], NOHEL, PEGO & TZAVARAS [131] the following mathematical model is considered for Poiselle-like shear flows of a non-Newtonian viscoelastic fluid between two parallel plates

$$u_t = S_x, \ (x,t) \in [0,1] \times [0,\infty),$$

$$\sigma_t + \sigma = g(u_x), \ (x,t) \in [0,1] \times [0,\infty),$$

where u is a velocity, $S = T + fx$, $T = \sigma + u_x$ is a shear stress, f is a constant gradient of pressure, and $g\colon \mathbb{R} \longrightarrow \mathbb{R}$ is a smooth odd function such that $p \cdot g(p) > 0$ for $p \neq 0$. It is easy to see that stationary solutions of this system are subject to the equation

$$[g(u_x) + u_x]_x = 0$$

($\sigma = g(u_x)$). In the case of non-monotone function $\omega\colon p \mapsto g(p)+p$ this equation is a stationary part of the equation (2). Such an ω appears in experiments; see, f.e., BELKIN, VINOGRADOV & LEONOV [7]. This system allows to simulate flows with flattenings and breaks of velocity profiles (apropos, flattening of the Poiselle profile top is characteristic for turbulent flows). An analogous model with $\omega(p) = p/(1+p^2) + \varepsilon p$ is discussed in OLDROYD [134] and JOHNSON & SEGALMAN [119] for modeling of polymer solutions flows.

1.7. Other applications

We shall describe briefly a series of applied problems in which an effect of variable sign viscosity appears.

ARSENJEV, GUBAR', RYKUNOV & SHELKOVNIKOV [1] consider equations, containing a diffusion coefficient of variable sign for investigation of vertical structure of turbulent tidal flows in rivers' estuary. These equations are received as a Rietz projection of averaged in a vertical component Euler-Reinolds equations.

The instability drift waves in non-isotermic non-uniform (in density) plasma in a magnetic field is explained in CHECHKIN, TUR & YANOVSKY [85] by alternating of viscosity sign within the framework of the Hasegawa-Mima model.

The equation similar to (8) is mentioned in HAZEWINKEL [113] in connection with the investigations of autocatalytic reactions. A detailed discussion see in HAZEWINKEL, KAASHOEK, & LEYNSE [115].

A Cahn-Hilliard type equations are suggested in COHEN & MURRAY [87] for description of diffusion processes in biological populations. MURRAY [128] derives an analogous equation for modeling of mammals skins coloring pattern formation.

In KOGAN & MOLEVICH [21] and KOGAN, MOLEVICH & ORAJEVSKII [22] equations with viscosity of variable sign were discussed in connection with simulation of acoustic waves in non-equilibrium molecular gases.

In YANENKO, KURBATKIN, KRUPCHATNIKOV & EIKHER [50] a model of circulation of a planet atmosphere is derived causing to FBPE.

2. Preliminaries

In this Section we discuss formulation of some BVPs for our equations, describe *a priori* estimates and notions of solutions.

2.1. Main assumptions on the function ω

We shall consider the equations (1) and (2) and *always assume* that the function $\omega: \mathbb{R} \longrightarrow \mathbb{R}$ is a sufficiently smooth function (unless otherwise specially stated) satisfying the following three assumptions:

$$\begin{gathered}\text{there exist } p_1, p_2 \in \mathbb{R} \text{ such that } \omega'(p) < 0 \text{ for } p \in (p_1, p_2) \\ \text{and } \omega'(p) > 0 \text{ for } p \in (-\infty, p_1) \cup (p_2, +\infty);\end{gathered} \tag{10}$$

$$\omega'(p) \geq d \text{ for some } d > 0 \text{ and all sufficiently large } |p|; \tag{11}$$

$$\omega(0) = 0. \tag{12}$$

We shall refer them as *main conditions.*

In most cases it is easy to generalize the results for case when the function ω has not one but many intervals of monotone decreasing.

In a series of statements it will be assumed that ω satisfies the additional condition

$$p\omega(p) \geq cp^2$$

for certain $c > 0$ and for all $p \in \mathbb{R}$.

Such functions ω we shall call *coercive.*

Everywhere below we shall denote by c_1 and c_2 the numbers $\omega(p_1)$ and $\omega(p_2)$, by p_0 the root of the equation $\omega(p) = c_2$ different from p_2, and by p_3 the root of equation $\omega(p) = c_1$ different from p_1 (see Fig. 2).

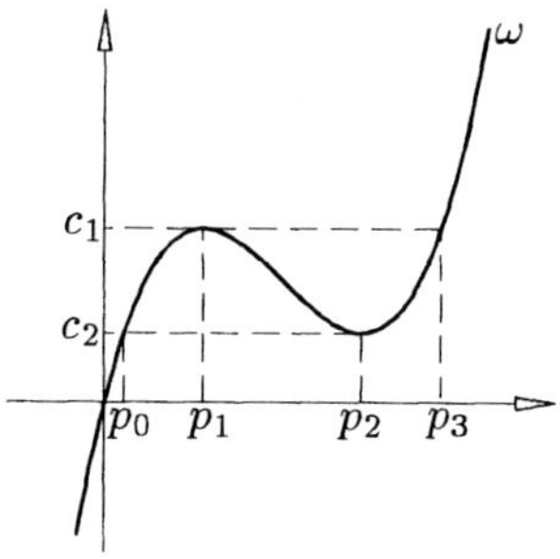

Fig. 2

2.2. Boundary value problems

The main attention in the review is given to the following two BVPs for equations (1) and (2), namely, to *Dirichlet BVP* on a function u: $\mathcal{D} \stackrel{\text{def}}{=} [0, 1] \times [0, T] \longrightarrow \mathbb{R}$ satisfying the equation in one sense or another, and satisfying the initial and boundary value conditions

$$u(x, 0) = u_0(x), \;\; x \in [0, 1]; \tag{13}$$

$$u(0, t) = a(t), \;\; u(1, t) = b(t), \;\; t \in [0, T], \tag{14}$$

and to *Neumann BVP*, in which the conditions (14) are replaced by boundary value conditions

$$u_x(0, t) = \alpha(t), \;\; u_x(1, t) = \beta(t), \;\; t \in [0, T]. \tag{15}$$

In these problems $u_0 \colon [0, 1] \longrightarrow \mathbb{R}$ and $a, b, \alpha, \beta \colon [0, T] \longrightarrow \mathbb{R}$ are given functions.

2.3. A minimum-maximum principle (Höllig & Nohel [118])

Let u is a H_1-solution of BVP (2), (13), (15) with $\alpha = \beta \equiv 0$. Then for all $(x, t) \in \mathcal{D}$

$$\min\{p_1, \min\{u_0'(x) \colon x \in [0, 1]\}\} \leq u_x(x, t) \leq$$
$$\leq \max\{p_2, \max\{u_0'(x) | x \in [0, 1]\}\}. \tag{16}$$

Obviously, the function $v = u_x$ is a solution of BVP (3), (13), (14) with zero a and b and initial function u_0'. We shall denote by M_2 the right-hand part of (16). Let $\varepsilon > 0$, $b_2(p) = (p - M_2 - \varepsilon)_+$, where $a_+ \stackrel{\text{def}}{=} \max\{0, a\}$), and

$B_2(p) = \frac{1}{2}[b_2(p)]^2$ (evidently, $B_2' = b_2$). Multiplying the identity $v_t = [\omega(v)]_{xx}$ by $b_2(v)$ and integrating them in x and t, after simple transformations we have

$$\int_0^1 B_2[v(x,t)]\,dx = -\int_0^t \int_0^1 \omega'[v(x,t)]b_2'[v(x,t)]v_x^2(x,t)\,dx\,dt +$$
$$+\int_0^1 B_2[u_0'(x)]\,dx \le 0,$$

or

$$\frac{1}{2}\int_0^1 [v(x,t) - M_2 - \varepsilon]_+^2\,dx \le 0.$$

By virtue of arbitrariness of ε, the latter means that

$$u_x(x,t) = v(x,t) \le M_2$$

and the right-hand part of (16) is proved. The proof of the left-hand part is completely similar.

For smooth solutions it is possible to prove a little more general statement (see NOVIKOV [33]).

Theorem. *Let u be a $C^{2,2}$-solution of the equation* (2), *and $\Gamma = \partial\mathcal{D}\backslash(0,1)\times\{T\}$ (the boundary of the cylinder $\mathcal{D}$ without the top cover). Then for all $(x,t) \in \mathcal{D}$*

$$\min\{p_1, \min\{u_x(x,t): (x,t) \in \Gamma\}\} \le u_x(x,t) \le$$
$$\le \max\{p_2, \max\{u_x(x,t): x \in \Gamma\}\}.$$

2.4. A priori estimates for the equation (2)

First, we shall assume that the function ω is coercive. In a standard way (multiplying by u and integration over $\mathcal{D}$) one can show that if $u \in H_1(\mathcal{D})$ is a solution of BVP either (2), (13), (14) or (2), (13), (15) with zero boundary conditions, then (see HÖLLIG & NOHEL [118], and also, BEREZIN, DUDNIKOVA, NOVIKOV & YANENKO [8])

$$||u_x||_{L_2(\mathcal{D})} \le \frac{1}{\sqrt{2c}}||u_0||_{L_2[0,1]}, \tag{17}$$

$$||u(\cdot,t)||_{L_2[0,1]} \le ||u_0||_{L_2[0,1]} \text{ for almost all } t \in [0,T].$$

Multiplication by u_t and integration results to the estimate

$$||u_t||_{L_2(\mathcal{D})} \le \left(\int_0^1 \Omega[u_0'(x)]\,dx\right)^{1/2}, \tag{18}$$

where $\Omega(p) \overset{\text{def}}{=} \int\limits_0^p \omega(s)\,ds$, and

$$\|u_x(\cdot,t)\|_{L_2[0,1]} \le \sqrt{\frac{2}{c}}\left(\int\limits_0^1 \Omega[u_0'(x)]\,dx\right)^{1/2} \quad \text{for almost all } t \in [0,T]. \tag{19}$$

If u is a solution of homogeneous Dirichlet problem then from (17) and Poincaré inequality the estimate

$$\|u\|_{L_2(\mathcal{D})} \le \frac{1}{2\sqrt{c}}\|u_0\|_{L_2[0,1]}$$

follows.

A little more complex (though standard) reasoning (J. Bona, J. Nohel, L. Wahlbin, see Höllig & Nohel [118], cf. also Lar'kin, Novikov & Yanenko [25]) allows to estimate $\|u_x\|_{L_\infty(\mathcal{D})}$. For any $n \in \mathbf{N}$ we define the function Ω_n assuming $\Omega_n(p) = \int\limits_0^p \omega^{2n+1}(s)\,ds$. Obviously, because of coercivity of ω,

$$\Omega_n(p) \ge \frac{c^{2n+1}}{2n+2}p^{2n+2}.$$

Multiplication of (2) by $\Omega_n(u_x)u_t$, integration over $\mathcal{D}$, and simple transformations leads to

$$\int\limits_0^t\int\limits_0^1 \omega^{2n+1}(u_x)u_t^2\,dx\,ds = \frac{1}{2n+1}\left[\int\limits_0^1 \Omega_n(u_0')\,dx - \int\limits_0^1 \Omega_n[u_x(x,t)]\,dx\right]. \tag{20}$$

As far as

$$\int\limits_0^1 \Omega_n[u_x(x,t)]\,dx \ge \frac{c^{2n+1}}{2n+2}\int\limits_0^1 u_x^{2n+2}(x,t)\,dx,$$

$$\int\limits_0^1 \Omega_n(u_0')\,dx \le \ |u_0'\|_{L_\infty[0,1]} \cdot \|\omega(u_0')\|^{2n+1}_{L_\infty[0,1]},$$

and the left-hand part of (20) is non-negative, then

$$\frac{c^{\frac{2n+1}{2n+2}}}{(2n+2)^{\frac{1}{2n+2}}}\|u_x(\cdot,t)\|_{L_{2n+2}[0,1]} \le \ |u_0'\|^{\frac{1}{2n+2}}_{L_\infty[0,1]} \cdot \|\omega(u_0')\|^{\frac{2n+1}{2n+2}}_{L_\infty[0,1]}.$$

Passing to the limit as n tends to ∞ implies the estimate

$$\|u_x\|_{L_\infty(\mathcal{D})} \le \frac{1}{c}\|\omega(u_0')\|_{L_\infty[0,1]}. \tag{21}$$

It is easy to give an example of absence of *a priori* estimates of u_{xx} for solutions from $H_1(\mathcal{D})$.

In all above estimates it is possible to weaken the coercivity condition: it is required only that $|\omega'(p)| \geq d$ for $|p|$ sufficiently large. Certainly, constants in the right-hand side of the estimates will be changed (see YANENKO & NOVIKOV [51], ZELENJAK [15], ZELENJAK, NOVIKOV & YANENKO [17], LAR'KIN, NOVIKOV & YANENKO [25]).

2.5. A priori estimates for the equation (1)

The presence of a convective term in the equations does not bring in principal difficulties in derivation of *a priori* estimates comparing with the equation (2), though overloads the proof. We only list the estimates (YANENKO & NOVIKOV [51], ZELENJAK [15], ZELENJAK, NOVIKOV & YANENKO [17], BEREZIN, DUDNIKOVA, NOVIKOV & YANENKO [8]) omitting the proofs, as far as they are described in detail in the monograph LAR'KIN, NOVIKOV & YANENKO [25]. In these estimates C is a constant dependent from ω, u_0 and T. So, for any solution of BVP (1), (13), (14) from $H_1(\mathcal{D})$ with zero boundary conditions the following estimates takes place

$$||u(\cdot,t)||_{L_2[0,1]} \leq C \text{ for almost all } t \in [0,T],$$

$$||u_x(\cdot,t)||_{L_2[0,1]} \leq C \text{ for almost all } t \in [0,T],$$

$$||u_t||_{L_2(\mathcal{D})} \leq C,$$

$$||u_x||_{L_\infty(\mathcal{D})} \leq C,$$

$$||[\omega(u_x)]_x||_{L_2(\mathcal{D})} \leq C.$$

An analogous estimate one can obtain for Neumann BVP.

As for the equation (2), it is easy to show impossibility of *a priori* estimation of u_{xx}.

2.6. A priori estimate for the second derivative of smooth solutions

In the case of three times continuously differentiable inside $\mathcal{D}$ solutions one can obtain *a priori* estimates for u_{xx} and u_{xt} that are typical for non-degenerate problems. Their derivation is based on an analogy of Lyapunov functional for parabolic equations (see ZELENJAK [16]). It is succeed to write out for solutions a relation of the form

$$\frac{d}{dt} f[u(\cdot,t)] = -F[u(\cdot,t)], \tag{22}$$

where functional F (need not be of fixed sign) appears to be of fixed sign on smooth solutions. We shall describe in detail this statement for the simplest homogeneous Dirichlet BVP for the equation (2) (see LAVRENTJEV, JR. [29]).

Theorem. *Let ω be a cubic polynomial (satisfying the main conditions), and u be a three times continuously differentiable inside $\mathcal{D}$ solution of the problem* (2), (13), (14) *with $a = b = 0$. Then for some C and any $\varepsilon \in (0, T]$*

$$\|u_{xt}\|_{L_2(\mathcal{D}_{T-\varepsilon})} + \|u_{xx}\|_{L_2(\mathcal{D}_{T-\varepsilon})} \leq C\varepsilon^{-1}. \tag{23}$$

Smoothness of the solution allows to assert that in $\mathcal{D}$

$$u_t \equiv \omega'(u_x)u_{xx}, \tag{24}$$

$$u_{xt} \equiv \omega''(u_x)u_{xx}^2 + \omega'(u_x)u_{xxx}, \tag{25}$$

$$u_t(0,t) \equiv \omega'[u_x(0,t)]u_{xx}(0,t) \equiv u_t(1,t) \equiv \omega'[u_x(1,t)]u_{xx}(0,t) \equiv 0. \tag{26}$$

These identities allow to write the relation (22) for a smooth solution of the equation (2) in the form (we differentiate $\int_0^1 \omega'(u_x)u_{xx}^2\,dx$ in t, integrate by parts with regard for (26), and use (24) and (25))

$$\frac{d}{dt}\int_0^1 \omega'(u_x)u_{xx}^2\,dx = -\int_0^1 \left(u_{xt}^2 + [\omega'(u_x)u_{xxx}]^2\right) dx -$$

$$-\int_0^1 \omega'(u_x)\omega''(u_x)u_{xx}^2 u_{xxx}\,dx. \tag{27}$$

Double integration by parts of the last term in (27), integration in t and elementary transformations result the equality

$$\int_0^1 [\omega'(u_x)u_{xx}^2]\Big|_{t=T} dx - \int_0^1 \omega'(u_0')(u_0'')^2\,dx =$$

$$= -\int_0^T\int_0^1 \left(u_{xt}^2 + [\omega'(u_x)u_{xxx}]^2\right) dx\,dt +$$

$$+\frac{1}{3}\int_0^T\int_0^1 [\omega''(u_x)u_{xx}^2]^2\,dx\,dt + \frac{1}{3}\int_0^T\int_0^1 \omega'(u_x)\omega'''(u_x)u_{xx}^4\,dx\,dt. \tag{28}$$

Further, hence ω is a cubic polynomial with the derivative of variable sign, then for any $C > 0$, one can choose $\delta > 0$ such that for any $|p| \leq C$

$$\omega'(p)\omega''(p) \geq \left(\frac{1}{2} - \delta\right)[\omega''(p)]^2. \tag{29}$$

Because of *a priori* estimate for u_x in $L_\infty(\mathcal{D})$ (see (21)),

$$\omega'(u_x)\omega''(u_x) \leq \left(\frac{1}{2} - \delta\right) [\omega''(u_x)]^2$$

for some $\delta > 0$. Therefore, continuing (28), we have

$$\int_0^1 [\omega'(u_x)u_{xx}^2]\big|_{t=\tau}\, dx \leq C_1 - \int_0^\tau\int_0^1 (u_{xt}^2 + [\omega'(u_x)u_{xxx}]^2)\, dx\, dt +$$

$$+\left(\frac{1}{2} - \frac{\delta}{3}\right)\int_0^\tau\int_0^1 [\omega''(u_x)u_{xx}^2]^2\, dx\, dt, \tag{30}$$

(where $C_1 = \int_0^1 \omega'(u_0')(u_0'')^2\, dx$). Further, (25) implies

$$\int_0^\tau\int_0^1 [\omega''(u_x)u_{xx}^2]^2\, dx\, dt \leq 2\int_0^\tau\int_0^1 (u_{xt}^2 + [\omega'(u_x)u_{xxx}]^2)\, dx\, dt, \tag{31}$$

and (24) and (19) imply

$$\int_0^\tau\int_0^1 \omega'(u_x)u_{xx}^2\, dx\, dt \leq C_2. \tag{32}$$

The inequalities (30)–(32) imply the estimate

$$\int_0^T\int_0^\tau\int_0^1 (u_{xt}^2 + [\omega'(u_x)u_{xxx}]^2)\, dx\, dt\, d\tau \leq C_3. \tag{33}$$

The well-known representation $\int_0^T\int_0^\tau \psi(t)\, dt\, d\tau = \int_0^T (T-t)\psi(t)\, dt$ leads to the inequality

$$\int_0^{T-\varepsilon}\int_0^1 (u_{xt}^2 + [\omega'(u_x)u_{xxx}]^2)\, dx\, dt \leq C_3\varepsilon^{-1}.$$

Now (25) and (equivalent to (18)) estimate $\|\omega'(u_x)u_{xx}\|_{L_2(\mathcal{D})}$ imply the inequality

$$\int_0^{T-\varepsilon}\int_0^1 ([\omega'(u_x)u_{xx}]^2 + [\omega''(u_x)u_{xx}^2]^2)\, dx\, dt \leq C_4\varepsilon^{-1}. \tag{34}$$

As far as ω' has no multiple roots, we notice now that

$$\text{there exists } \eta > 0 \text{ such that } (|\omega''(p)| < \eta) \Rightarrow (|\omega'(p)| > \eta). \tag{35}$$

Therefore, as it easy to see, that (34) implies the estimate

$$\eta^2 \int_0^{T-\varepsilon} \int_0^1 u_{xx}^2 \, dx \, dt - \eta^2 T \leq C_4 \varepsilon^{-1}.$$

The statement of the theorem follows now from (33) and the latter inequality.

We note that the proof can be generalized without changes for the case of any function $\omega \in C^3$ satisfying the main conditions and the conditions (29), (35).

The similar estimates take place for the smooth solutions of the equation (1) (see LAVRENTJEV, JR. [30]).

2.7. On unsolvability in a class of smooth solutions

Let us formally remove the external parenthesis in the equation (2), i.e., rewrite it in the form

$$u_t = \omega'(u_x) u_{xx}. \tag{36}$$

Obviously, in those points of $\mathcal{D}$ in which $\omega'(u_x) > 0$ on a given solution the equation (2) is parabolic, and in those points in which $\omega'(u_x) < 0$ it is backward parabolic one (also on a given solution).

In general case, an alternation of parabolicity direction results to unsolvability of classical BVPs for such equations. One can explain this fact easily in the case when the function ω is linear on (p_1, p_2) (see HÖLLIG & NOHEL [118], HÖLLIG [117]). Let, for definiteness, $\omega'(p) = d_1 < 0$ for $p \in (p_1, p_2)$. It is easy to prove that, for example, Dirichlet BVP (2), (13), (14) has no solutions u such that $u_x \in C(\mathcal{D})$ if the initial function u_0 on some segment $[x_1, x_2]$ is not analytic and "lies in the region of backward parabolicity", i.e., $u_0'(x) \in (p_1, p_2)$. In fact, if u is such a solution, by virtue of the continuity u_x, it follows that for T_1 sufficiently small the function $v(x,t) = u(x, T_1 - t)$ is a solution of parabolic problem

$$v_t = -d_1 v_{xx}, \ t \in [x_1, x_2] \times [0, T_1],$$

$$v(x, 0) = u(x, T_1), \ x \in [x_1, x_2],$$

$$v(x_1, t) = u(x_1, T_1 - t), \ v(x_2, t) = u(x_2, T_1 - t), \ t \in [0, T_1].$$

By virtue of highly smoothing property of the solution operator for parabolic problems (see, f.e., BERS, JOHN & SCHECHTER [78], p. 102), the function $x \mapsto v(x, T_1)$ will be analytic on $[x_1, x_2]$. And, as far as $v(x, T_1) \equiv u_0(x)$, this contradicts to non-analycity of u_0.

Absence of global smooth solutions for the problem (2), (13), (15) in the case when $u_0'(x) \in [p_1, p_2]$ for all $x \in [0, 1]$ (non-analycity of u_0 is not assumed) and ω is a cubic polynomial, is proved in NOVIKOV [33]. The proof is based on growth estimates of solutions in the region of backward parabolicity. Such estimates result to a contradiction with minimum-maximum principle.

3. Selfsimilar boundary value problems

Self-similar problems (for example, stationary one or the problems on traveling waves) are essentially more simple and, consequently, more advanced results for them are obtained.

3.1. Stationary solutions of a degenerate Cahn-Hilliard equation in variational posing

Let us consider degenerate ($\varepsilon = 0$) stationary CHE

$$[\omega(c)]_{xx} = 0.$$

This equation arises in description of steady states of two-phase solutions (see 1.2.), namely in a problem of minimization of the energy integral

$$\mathcal{F}_0(c) \stackrel{\text{def}}{=} \int_0^1 W[c(x)]\,dx \to \min; \tag{37}$$

integral $\mathcal{F}_0$ is a Gibbs free energy ($W' = \omega$). Let us supply the problem (37) by restrictions

$$\int_0^1 c(x)\,dx = m, \tag{38}$$

$$c'(0) = c'(1) = 0 \tag{39}$$

(the conservation of mass law and zero mass flux condition on the boundary). Here we assume that the function ω satisfies the conditions described in 1.2.. The minimization problem (37) with restriction (38) can be reduced to the problem on extremum of the functional

$$\int_0^1 \{W[c(x)] - \lambda c(x)\}\,dx,$$

where λ is a Lagrange multiplier. The Euler-Lagrange equation for it is

$$\omega(c) = \lambda,$$

and the angular Weierstrass-Erdmann condition is a continuity condition for the function $x \mapsto W[c(x)] - \lambda c(x)$ in discontinuity points of c. Simple analysis (see Carr, Gurtin & Slemrod [84]) shows that solutions of the problem (37)–(39) are piece-wise constant functions

$$c(x) = \begin{cases} c_A \text{ if } x \in \mathcal{D}_A, \\ c_B \text{ if } x \in \mathcal{D}_B, \end{cases} \tag{40}$$

where $\mathcal{D}_A$ and $\mathcal{D}_B$ are measurable subsets of the segment $[0,1]$ such that $\mathcal{D}_A \cap \mathcal{D}_B = \emptyset$, $\mathcal{D}_A \cup \mathcal{D}_B = [0,1]$, and numbers $c_A < c_B$ are determined by the following conditions

$$W(c_B) - W(c_A) = \lambda_0 (c_B - c_A), \quad \lambda_0 = W'(c_A) = W'(c_B);$$

(this means that a line parallel to the line $\{(c, \lambda_0 c)\}$ tangents the graph of W in two points; existence such a support line follows from "double-well" shape of W (see Fig. 3). From (38) follows that

$$C_A \operatorname{mes} \mathcal{D}_A + c_B \operatorname{mes} \mathcal{D}_B = m$$

and, besides,

$$\operatorname{mes} \mathcal{D}_A + \operatorname{mes} \mathcal{D}_B = 1.$$

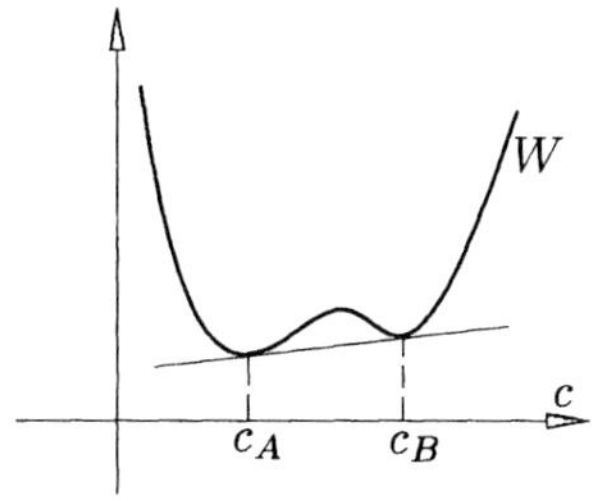

Fig. 3

Therefore, if $c_A < m < c_B$, then

$$\operatorname{mes} \mathcal{D}_A = \frac{m - c_B}{c_A - c_B}, \quad \operatorname{mes} \mathcal{D}_B = \frac{m - c_A}{c_B - c_A}. \tag{41}$$

The conditions (40) and (41) completely define a continuum of solutions of the problem (37)–(39).

In sec.5. a question on physical preference of these solutions will be discussed. It turns out that such solutions are functions (40) (we shall call them *simple solutions*) such that

$$\mathcal{D}_A = \frac{m - c_B}{c_A - c_B}, \text{ or } \mathcal{D}_B = \frac{m - c_B}{c_A - c_B}.$$

Many-dimensional variant of this variational problem is considered, in particular, in Gurtin & Matano, [112]. A geometrical structure of local minima of full energy in the various regions (cylinders, balls, spherical shells, disks, rings) is investigated.

In Gurtin [111] a short review of results concerning a problem of minimization of energy functional $\mathcal{F}_\varepsilon(c)$ is given, as well as, a series of unsolved problems and conjectures is formulated.

In the papers Sternberg [145], Kohn & Sternberg [121], Owen, Rubinstein & Sternberg [135] a problem on minimization of functional $\mathcal{F}_\varepsilon(c)$ in a context of singularly perturbed variational problems is discussed. With the help of methods based on concept of Γ-convergence of functionals given by De-Goirgi, and matched asymptotic expansions, asymptotics as $\varepsilon \to 0$ of extrema of this functional is discussed.

3.2. Stationary Dirichlet problem for the equation (1)

Apparently, the simplest among problems considered is Dirichlet BVP for stationary version of the equation (1), i.e., BVP

$$uu' = [\omega(u')]', \ x \in [0,1], \tag{42}$$

$$u(0) = a, \ u(1) = b; \tag{43}$$

where $u: [0,1] \longrightarrow \mathbb{R}$, and the prime denotes the differentiation with respect to x. This problem admits reasonably complete analysis, in particular, from our point of view, this analysis allows to formulate plausible conjectures on structure and behavior of solutions for non-stationary equations.

Single integration of (42) results to family of the first order ordinary differential equations not resolved with respect to the highest derivative

$$\omega(u') = \frac{1}{2}u^2 - c \stackrel{\text{def}}{=} g_c(u), \ x \in [0,1]. \tag{44}$$

It can be considered as *a differential inclusion* (see, f.e., FILIPPOV [46])

$$u' \in \omega^{-1}[g_c(u)], \ x \in [0,1]. \tag{45}$$

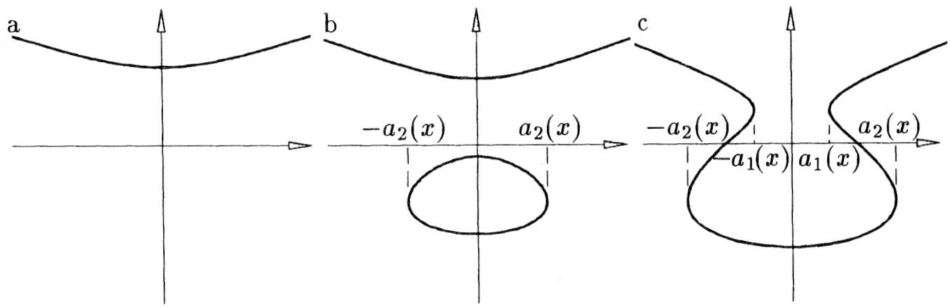

Fig. 4

Obviously, the function ω^{-1} has three maximal continuous branches f_1, f_2, f_3; let, for definiteness, they are numbered in such a way that on their common domain the inequalities

$$f_1(p) \geq f_2(p) \geq f_3(p)$$

take place. Evidently, for $c > c_1$ the function $\omega^{-1} \circ g_c$ is a single-valued function equals to $f_1 \circ g_c$. For $c_2 < c < c_1$ the function $\omega^{-1} \circ g_c$ on the interval $(-a_1(c), a_1(c))$, where $a_1(c) = \sqrt{2(c_1 - c)}$, is a three-valued function, and on $(-\infty, -a_1(c)) \cup (a_1(c), +\infty)$ is a single-valued one. Finally, for $c < c_1$ on $(-a_1(c), -a_2(c)) \cup (a_2(c), a_1(c))$, where $a_2(c) = \sqrt{2(c_2 - c)}$, it is three-valued, and on $(-\infty, -a_1(c)) \cup (-a_2(c), a_2(c)) \cup (a_1(c), +\infty)$ is single-valued (see Fig. 4).

One can consider that in some special sense the equations (44) and (45) are equivalent to the set of three ordinary differential equations in the normal form

$$u' = f_i[g_c(u)] \ (i = 1,2,3). \tag{45_i}$$

We shall say that absolutely continuous function $u: [0,1] \longrightarrow \mathbb{R}$ is a *generalized solution* of the equation (42), if for some c it turns the equation (44) into an identity almost everywhere on $[0, X]$. It is easy to show that such a solution is a generalized solution in usual for elliptic equations sense.

Piece-wise twice continuously differentiable generalized solution of the equation (42) we shall call *a quasigeneralized solution*. We introduce also one more special class of solutions, which, roughly speaking, can change continuous branches of the mapping $\omega^{-1} \circ g_c$ if and only if $u(x)$ at a given point x lies on the discriminant curve of the equation (44). Namely, we shall call quasigeneralized solution of the equation (42) *strong* one, if in any point x_0 of discontinuity of u' the number $g_c[u(x_0)]$ is equal to either c_1, or c_2. Finally, by *a classical solution* of the equation we shall mean a function $u \in C^2$, satisfying the equation everywhere on $[0,1]$. Let us denote by C_{ab}, S_{ab}, Q_{ab} and G_{ab} sets of classical, strong, quasigeneralized and generalized solutions of the problem (42)–(43), respectively. Obviously, $C_{ab} \subset S_{ab} \subset Q_{ab} \subset G_{ab}$.

3.3. Solvability of the problem (42)–(43)

It is easy to specify values of parameters a and b for which the problem (42)–(43) has no classical solutions. The measure of the set of such parameters in $\mathbb{R}^2$ is infinite.

The following theorems can be easily (though clumsily) proved by a shooting method with shooting parameter c (AKHMEROV & NOVIKOV [5,6]).

Theorem. *The sets Q_{ab} and G_{ab} for any $a, b \in \mathbb{R}$ are either single-point, or continual.*

Theorem. *If $0 \notin [p_1, p_2]$, then the set S_{ab} for any $a, b \in \mathbb{R}$ is nonempty and not more than countable. On a set of parameters (a, b) of a nonzero measure it is countable. If $0 \in (p_1, p_2)$, on some set of parameters (a, b) of a nonzero measure the set S_{ab} is empty.*

We omit the proofs of these statements.

The theorems on a solvability of stationary Neumann BVP for the equation (42) in various classes of solutions were proved in AKHMEROV [3].

3.4. Compactness properties of solutions set for the problem (42)–(43)

The estimate

$$\|u\|_{C^1} \leq C$$

(it can be proved just in the same way as in non-stationary case (see Subsection 2.5.) implies compactness of the set G_{ab} in the space C for any $a, b \in \mathbb{R}$.

Theorem (AKHMEROV [53,54]). *If $0 \in (p_1, p_3)$ then the set S_{ab} is either finite, or not relatively compact in the spaces W_p^1 $(1 \leq p \leq \infty)$.*

We show here only noncompactness of the set S_{00}. For this purpose we present a sequence $\{u_k\}$ of strong solutions of (42)–(43) such that $\|u_k - u_l\|_{W_p^1} \geq \alpha$ for $k \neq l$. For any c in a small left-hand neighborhood of c_1 (in this case $\omega^{-1} \circ g_c$ has a shape represented on Fig. 4b) we denote by u_c the strong solution of the equation (45) satisfying the following conditions:

(i) $u_c(0) = 0$;

(ii) u_c has a countable number of points $x_n(c)$ $(n = 1, 2, \ldots)$ of discontinuity of the derivative numbered in ascending order, and also $u_c[x_n(c)]$ is equal to either $a_1(c)$ or $-a_1(c)$;

(iii) on the odd segments $[x_{2n-1}(c), x_{2n}(c)]$ the function u_c is a solution of the equation (45_1), and on the even one it satisfies the equation (45_3).

Let $z_n(c)$ are zeroes of the function u_c numbered in ascending order. The usual continuous dependence theorems (see, f.e., HARTMAN [47]; the uniqueness of solutions required in this theorem follows from the results by SADOVSKII [43]) imply continuity of the functions $c \mapsto z_n(c)$, moreover, $z_n(c) \longrightarrow 0$ as $c \to c_1 - 0$ for any n, and $z_n(c) \longrightarrow \infty$ as $n \to \infty$ for any c. Let now $n^* \in \mathbf{N}$ and $c^* \in (c_1 - \varepsilon, c_1)$ be such that $z_{2n^*}(c^*) = 1$, $n_k = 2^k n^*$ and c_k is a solution of the equation $z_{n_k}(c) = 1$ (its existence follows from the above mentioned properties of z_n). It is easy to show that the sequence $\{u_{c_k}\}$ the desired one.

Two remarks. First, it should noted that $\{u_k'\}$ are in some sense an analogy of the Rademacher system (see, f.e., ZYGMUND [18]). Second, one can show that the Hausdorff measure of noncompactness (see, f.e., AKHMEROV, KAMENSKII, POTAPOV, RODKINA & SADOVSKII [4]) $\chi(S_{00})$ of the set S_{00} in the space W_p^1 is equal to $(p_3 - p_1)^{\frac{p-1}{p}} \min\{p_3, p_1\}$.

Theorem. *The sets Q_{ab} and G_{ab} are either single-point, or not relatively compact in W_p^1 $(1 \le p \le \infty)$.*

In fact, let $u_0 \in Q_{ab}$, $c = \omega[u_0'(1) - \frac{1}{2}u_0^2(1)]$, and $U_c = \{u \in \mathbb{R}$: the set $\omega^{-1}[g_c(u)]$ is not single-point$\}$. If $u_0([0,1]) \cap U_c$ is empty or single-point, then from monotonicity reasons it follows that Q_{ab} is single-point. Otherwise, $u_0([0,1]) \cap U_c$ contains some segment I and one can construct not relatively compact sequence of quasigeneralized solutions $\{u_k\}$ coinciding with u_0 outside of $u_0^{-1}(I)$ and "jumping" from one branch of the mapping $\omega^{-1} \circ g_c$ on $u_0^{-1}(I)$ to another. Solutions u_k can be constructed in the way similar to described in the proof of the previous theorem.

Connectedness of sets of generalized and quasigeneralized solutions of the Cauchy problem for the equation (48) in the spaces $W_p^1[0,1]$ $(1 \le p < \infty)$ and their disconnectedness in the space $W_\infty^1[0,1]$ has been proved in AKHMEROV [2].

3.5. Closures of solutions set of the problem (42)–(43)

We shall denote closure of a set A in a Banach space E by $\mathrm{cl}_E A$. The following theorem (AKHMEROV [53,54]) is a simple corollary of the fact that any point from Q_{ab} and G_{ab} is isolated in W_∞^1 and any point from S_{ab} is isolated in W_p^1 $(1 \le p \le \infty)$.

Theorem. $\mathrm{cl}_{W_\infty^1} Q_{ab} = Q_{ab}$, $\mathrm{cl}_{W_\infty^1} G_{ab} = G_{ab}$, $\mathrm{cl}_{W_p^1} S_{ab} = S_{ab}$.

In the standard way one can prove also the

Theorem. $\mathrm{cl}_{W_p^1} Q_{ab} = \mathrm{cl}_{W_p^1} G_{ab} = G_{ab}$.

Closure of the sets Q_{ab} and G_{ab} in C turn out to be a set of solutions of a differential inclusion that is derived from (42) by a special convexification procedure. Let us define a multi-valued function $\Omega: \mathbb{R} \longrightarrow 2^{\mathbf{R}}$ as inverse to the multi-valued function

$$\Omega^{-1}(p) = [\min\{\omega^{-1}(p)\}, \max\{\omega^{-1}(p)\}]$$

(see Fig. 5). Along with the equation (42) we consider the differential inclusion

$$uu' \in [\Omega(u')]', \ x \in [0,1], \tag{47}$$

which we interpet as family of differential inclusions

$$u' \in \Omega^{-1}[g_c(u)], \ x \in [0,1]. \tag{48}$$

By *a solution of the inclusion* (47) we mean, as usually (see, f.e., FILIPPOV [46]), an absolutely continuous function satisfying (48) almost everywhere on $[0,1]$. We denote a set of solutions of BVP (47), (43) by I_{ab}.

Theorem [53,54]. $\mathrm{cl}_C Q_{ab} = \mathrm{cl}_C G_{ab} = I_{ab}$.

The inclusion $\mathrm{cl}_C Q_{ab} \subset \mathrm{cl}_C G_{ab}$ is evident, and the $\mathrm{cl}_C G_{ab} \subset I_{ab}$ can be easily proven. We note for the proof of the inclusion $I_{ab} \subset \mathrm{cl}_C Q_{ab}$ that for any solution u_0 of the inclusion (47) on any segment $[\alpha,\beta] \subset [0,1]$ there exists a quasigeneralized solution of the equation (42) such that it coincides with u_0 at the ends of segment $[\alpha,\beta]$. This quasigeneralized solution can be constructed out of solutions of the Cauchy problems

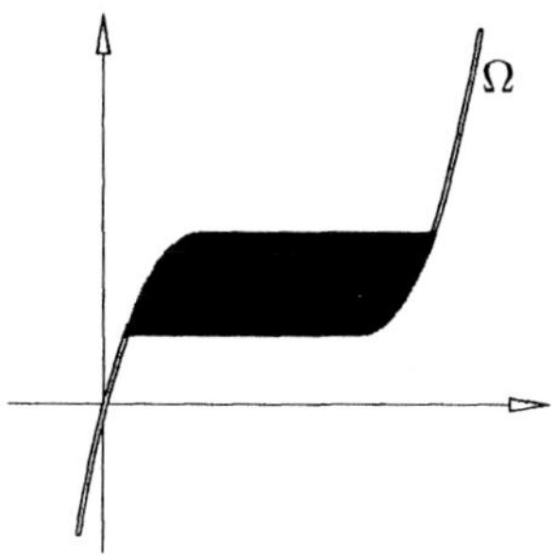

Fig. 5

$$u' = \omega_*^{-1}[g_c(u)], \ x \geq \alpha, \ u(\alpha) = u_0(\alpha),$$

and

$$u' = \omega_{**}^{-1}[g_c(u)], \ x \leq \beta, \ u(\beta) = u_0(\beta),$$

in which $\omega_* = \omega|_{(-\infty,p_0]\cup(p_2,+\infty)}$, and $\omega_{**} = \omega|_{(-\infty,p_1]\cup(p_3,+\infty)}$. Therefore, for any $u_0 \in I_{ab}$ we can construct a sequence $\{u_n\}$ of quasigeneralized solutions of the problem (42)–(43), coinciding with u_0 at the points i/n $(i = 0,1,\ldots)$ of the segment $[0,1]$. Obviously, this guarantees the wanted inclusion.

3.6. Traveling waves of the forward backward convection-diffusion equation

We shall say that a solution u of the equation (1) is *a traveling wave*, if it is of the form

$$u(x,t) = U(x - \alpha t), \tag{49}$$

where $U\colon \mathbb{R} \longrightarrow \mathbb{R}$, and $\alpha \in \mathbb{R}$ is a parameter (*velocity of traveling wave*). We shall consider three classes of traveling waves selected by the following boundary conditions

$$U \text{ is a periodic function;} \tag{50}$$

$$U(-\infty) = A, \ U(+\infty) = B \ (A > B); \tag{51}$$

$$u(-\infty) = A, \ U \text{ is asymptotically periodic on } +\infty; \tag{52}$$

where $U(\pm\infty) = \lim_{s\to\pm\infty} U(s)$, and asymptotic periodicity means that U can be represented as sum of a periodic function and a function tending to zero on

$+\infty$. Traveling waves of the equation (1) satisfying boundary conditions (50), (51) and (52) we shall call *periodic*, *shock transition on zero background* and *shock transition on oscillating background*, respectively. In the first and third problems the parameter α is given, and in the second one the wave's velocity is determined by the problem's parameters: $\alpha = (A - B)/2$.

As well as in above problems, in general case, alternation of ω' sign implies absence of smooth traveling waves. Therefore we shall consider only some subclasses of generalized solutions. Now we describe these classes. Let us substitute (49) in (1) and integrate with respect to x:

$$\omega(U') = \frac{1}{2}U^2 - \alpha U + C;$$

where the prime denotes the differentiation with respect to $s = x - \alpha t$, and C is a constant of integration. Let us consider new unknown function $v = U - \alpha$, denote $C - \alpha^2/2$ by c, and the mapping $v \mapsto v^2/2 + c$ by g_c; then we have the equation

$$\omega(v') = g_c(v). \tag{53}$$

In the problem (1), (50) the constant c is arbitrary, and in the problems (1), (51) and (1), (52) it is equal to $-(A - B)^2/8$ and $-(A - \alpha)^2/2$, respectively.

As well as in 3.2., for the equation (53) notions of strong, quasigeneralized and generalized solutions can be defined. These notions generate notions of strong, quasigeneralized and generalized traveling waves.

3.7. A structure of periodic traveling waves set for the equation (1)

We restrict ourselves only to formulations of results (see AKHMEROV [55,56]), as far as their proofs are reasonably close to the proof of similar statements concerning solutions set for stationary Dirichlet BVP (see 3.3., 3.4., 3.5.). We shall assume that $\omega \in C^3$, *and, moreover, if* $0 \in \{p_1, p_2\}$ *then* $\omega'(0) \neq 0$ *and* $\omega''(0) \neq 0$.

Obviously, the constant functions $u(x,t) \equiv \pm\sqrt{-2c}+\alpha$ are periodic traveling waves of the equation (1). Hereinafter we drop them from consideration and *call by periodic traveling waves only non-constant solutions.* Let us denote by s_c, q_s and g_s sets of non-constant strong, quasigeneralized and generalized periodic solutions of the equation (53) for given $c \in \mathbb{R}$, respectively, and by $S_{c,\alpha}$, $Q_{c,\alpha}$ and $G_{c,\alpha}$ corresponding sets of periodic traveling with velocity α waves of the equation (1) (correspondence between these sets establishes the mapping $v \mapsto u \equiv v(x - \alpha t + \alpha))$. Below c and α are fixed. Obviously, $S_{c,\alpha} \subset Q_{c,\alpha} \subset Q_{c,\alpha}$. We let $\Omega_+ = \omega([0,+\infty))$, $\Omega_- = \omega((-\infty,0])$, $\Omega_0 = \Omega_+ \cap \Omega_-$, and denote the interior of Ω_0 by $\operatorname{int}\Omega$.

Theorem. *If* $\int \Omega_0 \neq \emptyset$ *and* $c < \max \Omega_0$, *then* $Q_{c,\alpha} \neq \emptyset$. *Otherwise,* $G_{c,\alpha} = \emptyset$. *The set* $S_{c,\alpha}$ *is not empty (and, moreover, single-point) if and only if* $0 \in (p_1, p_2)$ *and* $c < p_1$.

Below we denote by Pc the set of all functions $u: \mathbb{R} \times \mathbb{R} \longrightarrow \mathbb{R}$ representable in the form (49) with continuous periodic function U. Let $p(U)$ be a minimal positive period of a function U ($p(\text{const}) \stackrel{\text{def}}{=} 0$). The set Pc becomes a complete metric space with respect to the metric $\rho_{Pc}(u,w) = \max\{|U(s) - W(s)|: s \in [0, p(U,W)]\} + |p(U) - p(W)|$, where $p(U,W) = \max\{p(U), p(W)\}$. Let us consider the equivalence relation $\mathcal{R}$ on Pc, assuming $U \mathcal{R} V$, if $U(s+\tau) \equiv V(s)$ for some $\tau \in \mathbb{R}$. The factor space of Pc with respect to $\mathcal{R}$ with the appropriate factor metric we denote by PC. We define also the set $Pw_p^1 \subset Pc$ of functions generated by formula (49) by absolutely continuous functions U with locally summable with pth degree derivative. The metric on Pw_p^1 we shall define by the formula $\rho_{Pw_p^1}(u,w) = \rho_{Pc}(u,w) + \left(\int\limits_0^{p(U,W)} |U'(s) - W'(s)|^p \, ds \right)^{1/p}$. The relation $\mathcal{R}$ results the factor space PW_p^1. Finally, the spaces Pw_∞^1 and PW_∞^1 can be defined similarly if we put $\rho_{Pw_\infty^1}(u,w) = \rho_{Pc}(u,w) + \text{ess sup}\{|U'(s) - W'(s)|: s \in [0, p(U,W)]\}$. Obviously, PW_p^1 $(1 \le p \le \infty)$ are complete metric spaces. Below by a function from PC or PW_p^1 we mean an appropriate equivalence class and the words "a unique solution" signifies uniqueness to within shifts.

Theorem. *The sets $Q_{c,\alpha}$ and $G_{c,\alpha}$ are relatively compact in PC.*

Theorem. *The sets $Q_{c,\alpha}$ and $G_{c,\alpha}$ are not relatively compact in PW_p^1 $(1 \le p \le \infty)$.*

Theorem. $\text{cl}_{PW_p^1} Q_{c,\alpha} = \text{cl}_{PW_p^1} G_{c,\alpha} = G_{c,\alpha}$ *for* $1 \le p < \infty$, $\text{cl}_{PW_\infty^1} Q_{c,\alpha} = Q_{c,\alpha}$, $\text{cl}_{PW_\infty^1} G_{c,\alpha} = G_{c,\alpha}$.

For description of closures of $Q_{c,\alpha}$ and $G_{c,\alpha}$ in PC we define (in almost the same way as in the case of stationary Dirichlet BVP) the multi-valued function $\Omega: \mathbb{R} \longrightarrow 2^{\mathbb{R}}$ giving its graph by equality

$$\text{Gr}\,\Omega = \text{Gr}\,\omega \cup \left\{(p,q): q \in \Omega_0,\ p \in [\min\{\omega^{-1}(q)\}, \max\{\omega^{-1}(q)\}]\right\}.$$

We shall consider the problem on periodic traveling waves for differential inclusion

$$u_t + uu_x \in [\Omega(u_x)]_x. \tag{54}$$

Certainly, the right-hand part of this inclusion requires an interpretation. Just in the same way as above, the problem on periodic traveling waves can be reduced to the problem on periodic solutions of ordinary differential inclusion

$$\Omega(v') \ni g_c(v),$$

or, that is the same,

$$v' \in \Omega^{-1}[g_c(v)]. \tag{55}$$

We shall denote by $I_{c,\alpha}$ a set of periodic traveling waves of (54) generated by (absolutely continuous) periodic solutions of (55).

Theorem. $\text{cl}_{PC} Q_{c,\alpha} = \text{cl}_{Pc} G_{c,\alpha} = I_{c,\alpha}$.

The similar statements take place for traveling shock transitions.

4. Existence and uniqueness of solutions

In this Section we present existence and uniqueness theorems proven at present for Dirichlet and Neumann BVPs for FBPE and also some results on asymptotic behavior of solutions.

4.1. Uniqueness theorem for smooth solutions

We continue here discussion of Neumann BVP for the forward backward heat equation (2), (13), (15). We shall call a function $u: \mathcal{D} \longrightarrow \mathbb{R}$ such that $\mathcal{D} = [0,1] \times [0,\infty)$, u, u_x, u_{xx}, u_t are continuous in $\mathcal{D}$, $u_{xt} \in L_{\mathrm{loc},\infty}(\mathcal{D})$, and u_{xt} is continuous almost everywhere on $\mathcal{D}$ *smooth global solution* of this problem.

Theorem (LAIR [123,124]). *Let $u_0 \in C^1[0,1]$, $\alpha, \beta \in C[0,\infty)$, and ω is coercive, piece-wise continuously differentiable, and has on any compact interval a finite number of extrema. Let ω' be non-zero almost everywhere. Then the problem* (2), (13), (15) *can not have more than one smooth global solution.*

We outline the proof in a less bulky case, when ω satisfies the main conditions.

Let u and v are smooth global solutions of this problem and $w = u - v$. The statement of the theorem will be proven, if we show that

$$W(t) = \|w(\cdot,t)\|^2_{L_2[0,1]} = 0.$$

First, notice that W is bounded. In fact, integration of the identity

$$W_t = [\omega(u_x) - \omega(v_x)]_x \tag{56}$$

with respect to x in view of the boundary conditions implies

$$\left[\int_0^1 w(x,t)\,dx\right]_t = 0,$$

for $t \geq 0$. The latter together with the initial condition causes the identity

$$\int_0^1 w(x,t)\,dx \equiv 0.$$

But then, the standard imbedding theorems (see, f.e., LADYŽENSKAJA, SOLONNIKOV & URAL'CEVA [24]) leads to the inequality

$$W(t) = \|w(\cdot,t)\|^2_{L_2[0,1]} \geq \|w_x(\cdot,t)\|^2_{L_2[0,1]}.$$

By virtue of the *a priori* estimate (see Subsection 2.4.), u_x and v_x, and with them w_x, lie in $L_\infty(\mathcal{D})$. This guarantees the boundedness of W.

Multiplication of (56) by w and integration in x give after simple transformations

$$W'(t) = -2 \int_0^1 w_x[\omega(u_x) - \omega(v_x)]\, dx.$$

We show that if $W(t_0) > 0$ for some t_0, then the function $W'(t)$ is strictly separated from zero for $t \geq t_0$; this will contradict to boundedness of W. The latter, in turn, follows from the fact that

$$w_x[\omega(u_x) - \omega(v_x)] \leq 0, \tag{57}$$

and, moreover,

$$w_x[\omega(u_x) - \omega(v_x)] \leq -c(\varepsilon) w_x^2 \tag{58}$$

in such points of $\mathcal{D}$ where $w_x^2 \geq \varepsilon$ ($c\colon (0,\infty) \longrightarrow (0,\infty)$). This is a main point of the proof. It can be proved in the following way. Let $p_0 = \inf\{\min\{u_x(x,t), v_x(x,t)\}\colon (x,t) \in \mathcal{D}\}$, $p_3 = \sup\{\max\{u_x(x,t), v_x(x,t)\}\colon (x,t) \in \mathcal{D}\}$, and $P_i = [p_{i-1}, p_i]$ ($i = 1,2,3$). Obviously, $\mathcal{D} = \bigcup_{i=1}^{3} \mathcal{D}_{ij}$, where $\mathcal{D}_{ij} = \{(x,t) \in \mathcal{D}\colon u_x(x,t) \in P_i,\ v_x(x,t) \in P_j\}$. We show that (57) and (58) are fulfilled in all $\mathcal{D}_{ij}$. Let, by definition, $[a]_i$ be a if $a \in P_i$, p_{i-1} if $a \leq p_{i-1}$, and p_i if $a \geq p_i$. Let us "decompose" solutions u and v into "parts" u^i and v^i ($i = 1,2,3$) which correspond to a monotonicity region of the function ω, in the following way

$$u_x^i(x,t) = [u_x(x,t)]_i,\ u_t^i(x,t) = \begin{cases} u_t(x,t) & \text{if } u_x(x,t) \in P_i, \\ 0 & \text{if } u_x(x,t) \notin P_i. \end{cases}$$

We assume that the functions u^i satisfy the boundary conditions (15) with the functions $\alpha^i(t) = [\alpha(t)]_i$, $\beta^i(t) = [\beta(t)]_i$, respectively, and the initial condition (13) with the function $u_0^i(x) = \int_0^x [u_0'(s)]_i\, ds$ (the functions v^i are defined similarly). It can be shown that the functions u^i and v^i satisfy (2). Besides, $u^1 \equiv v^1$, $u^3 \equiv v^3$, as far as the function ω' is non-negative on u_x^1, v_x^1, u_x^3, v_x^3 and one can use the uniqueness theorem for nonlinear parabolic equations. Therefore, the inequalities (57) and (58) in $\mathcal{D}_{11}$ and $\mathcal{D}_{33}$ are equalities. It is easy to see also that (58) in $\mathcal{D}_{22}$ takes place because $\omega'(p) < 0$ almost everywhere on P_2. Furthermore, $u_x = u_x^1$, $v_x = v_x^2$, $v_x^1 = p_1$ in $\mathcal{D}_{12}$ (the rest $\mathcal{D}_{ij}$ are analogous) and, besides, $u_x^1 = v_x^1 = p_1$, as far as $u^1 \equiv v^1$. Therefore, $u_x, v_x \in P_2$, and hence (58) is fulfilled because $\omega(p) < 0$ almost everywhere on P_2. This completes the proof.

If we require in addition that $\omega'(p) < 0$ for all $p \in [p_1, p_2]$ (where $\omega'(p_1)$ and $\omega'(p_2)$ mean right and left derivatives, respectively), it follows uniqueness for local smooth solutions (2), (13), (15). A proof in this case is simpler because of opportunity to prove that $u^2 \equiv v^2$. The latter comes from the fact that the function $u^2 - v^2$ is a solution of the homogeneous Neumann BVP for some linear strictly backward parabolic equation (for more details see Lair [123]).

A uniqueness theorem for generalized solutions of Dirichlet BVP for the forward backward convection-diffusion equation (1), (13), (14) with $a = b = 0$ in the case when the initial function u_0 lies in the parabolicity segment: $u_0'(x) \in [\tilde{p}_1, \tilde{p}_2] \subset (-\infty, p_3] \cup [p_4, +\infty)$ was proved in NOVIKOV [32].

4.2. Existence theorem in a special case

First, we give an existence theorem for Neumann BVP for the forward backward heat equation in one special case (HÖLLIG [117], HÖLLIG & NOHEL [118]).

Theorem. *Let ω be a linear function on intervals $(-\infty, p_1)$, (p_1, p_2) and (p_2, ∞), and is coercive. If $u_0 \in C^2$, $u_0'(0) = u_0'(1) = 0$, $\alpha = \beta = 0$ and $\{x \in [0,1]: u_0'(x) \in (p_1, p_2)\} \neq \emptyset$, then for some $T > 0$ the problem* (2), (13), (15) *has in $\mathcal{D}$ a continuum of solutions in a class $\{u: \omega(u_x) \in C^\gamma,\ u_x \in L^\infty,\ u_t \in L^p\}$, where C^γ is the Hölder space ($\gamma < 1$).*

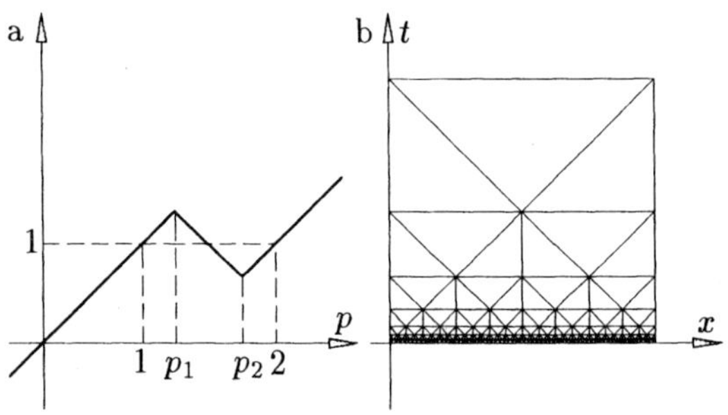

Fig. 6

The main idea of the proof consists of extracting of that part of a solution, which "jumps" from one branch of monotonicity of ω onto another one (not changing the value of $\omega(u_x)$). We describe it in an elementary situation, assuming that $\omega'(p) \equiv 1$ on $(-\infty, p_1) \cup (p_2, \infty)$. For the sake simplicity, we shall also consider that $1 < p_1 < p_2 < 2$, $\omega(p) \equiv 1$ on (p_1, p_2) and $u_0'(x) \leq 2$ for all $x \in [0,1]$ (see Fig. 6a).

Suppose we have constructed a function $w: \mathcal{D} \longrightarrow \mathbb{R}$ with the following properties: (i) $w_x \in \{0, 1\}$;

(ii) $w_t \in L^\infty$;

(iii) $\lim_{t\to 0} \|w(\cdot,t) - w_0\|_{L^\infty} = 0$, where $w_0(x) = \int\limits_0^x [u_0'(s) - 1]_+ ds$ (the index "+" means positive cutoff of a function: $[a]_+ = \max\{0, a\}$);

(iv) $\operatorname{supp}\{w_x(\cdot,t)\} \subset \{x \in [0,1]: u_0'(x) \geq 1\}$.

It is easy to see that the function $u = v + w$ is a solution of the problem (2), (13), (15), iff v is a solution of the problem

$$v_t = v_{xx} - w_t,\ (x,t) \in \mathcal{D}, \tag{59}$$

$$v(x,0) = u_0(x) - w_0(x),\ x \in [0,1], \tag{60}$$

$$v_x(0,t) = v_x(1,t) = 0. \tag{61}$$

This follows from the correlation $\omega(p) = \omega(p+1)$ (it holds for p sufficiently close to 1), condition (i), and closedness of v_x to 1 for t sufficiently small:

$$u_t = v_t + w_t = v_{xx} = [\omega(v_x)]_x = [\omega(v_x + w_x)]_x = [\omega(u_x)]_x. \tag{62}$$

Existence of a solution v of the problem (59), (60), (61) such that $v_x \in C_\alpha$ and $v_t \in L^p$ can be proved in a standard way (see HÖLLIG [117]).

It remains to show an existence of w, satisfying properties (i)–(iv). We shall describe here a construction, suggested by STRANG & ABDEL-NABY [146] (a little more complex idea was proposed by HÖLLIG [117]). Let

$$\mathcal{T} = \{y_{ij} = (j2^{-i}, 2^{-i}) \in \mathcal{D}: i \in \mathbf{N},\ j = 0, \ldots, 2^i\}$$

is a set of triangulation vertices of the area $\mathcal{D}$ (see Fig. 6b). The function w is defined as a piece-wise linear continuation onto $\mathcal{D}$ of the function preassigned on $\mathcal{T}$ by the relation

$$w(y_{ij}) \in \{k2^{-i}: k \in \mathbf{N}\},\ w(y_{ij}) \le w_0(j2^{-i}) \le w(y_{ij}) + 2^{-i}.$$

Obviously, such a function w satisfies the conditions (i)–(iv). It remains to notice that the function $(x,t) \mapsto w(x, \lambda t)$ satisfies these conditions for any $\lambda > 0$. For different λ we have distinct solutions of the problem (2),(13),(15).

It is easy to see also that the constructed solutions for all $(x,t) \in [0,1]\times(0,T]$ are such that $u_x \notin [p_1, p_2]$, i.e., $\omega'(u_x) > 0$.

4.3. On a structure of solutions set

In connection with a non-uniqueness of generalized solutions of the problem (2), (15), (13) the following question is of interest: how rich is a set of such solutions?. The particular answer to this question we give in the present subsection. We describe the statement in at most simple situation.

Theorem (AKHMEROV [57]). *Let the conditions of Subsection 4.2. are fulfilled and, besides, $p_1 > 0$, $\omega(p_2) < 0$. Let $u_0 \in C^2[0,1]$, $u_0'(0) = u_0'(1) = 0$ and $u_0'(x) \in [0,b]$ for all $x \in [0,1]$. Let $\varphi \in C^{2,1}(\mathcal{D})$, $\varphi(x,0) \equiv u_0(x)$ and $\varphi_x(x,t) \in [0,1]$ for all $(x,t) \in \mathcal{D}$. Let, finally, ψ be a solution of the following BVP:*

$$v_t = av_{xx} - \varphi_t(x,t),\ (x,t) \in \mathcal{D}, \tag{63}$$

$$v_x(0,t) = v_x(1,t) = 0, \tag{64}$$

$$v(x,t) = 0,\ x \in [0,1]. \tag{65}$$

Then for some $T > 0$ the function $\varphi + \psi$ lies in a closure of generalized solutions set of the problem (3)–(13) *in the space $C(\mathcal{D})$.*

In other words, if $\Phi = \{\varphi \in C^{2,1}(\mathcal{D}): \varphi$ satisfies the conditions (15), (13), $\varphi_x(x,t) \in [0,1]$ for all $x \in [0,1]\}$, $D_t = \frac{\partial}{\partial t}$, $\mathcal{L}$ is a differential operator of the problem (63)–(65), and $\mathrm{cl}_{C(\mathcal{D})}(S)$ is a closure of the set S of generalized solutions of the problem (3) in $C(\mathcal{D})$, (15), (13), then

$$(I + \mathcal{L}^{-1}D_t)\Phi \subset \mathrm{cl}_{C(\mathcal{D})}(S).$$

The main idea of the proof is a slightly modified G. Strang idea, described in Subsection 4.2.. We present a sequence of functions φ^n, approximating in

$C(\mathcal{D})$ a function φ, and a sequence ψ^n, approximating ψ. To construct φ^n we define a triangulation of area $\mathcal{D}$ by vertices $\mathcal{T}_n = (\mathcal{T}_n^0 \cup \mathcal{T}_n^1) \cap \mathcal{D}$, where

$$\mathcal{T}_n^0 = \{y_{ij}^0 = (2^{-n-i}j, 2^{-n-i}) \colon i = 0, 1, \ldots;\ j = 0, \ldots, 2^{n+i}\},$$

and

$$\mathcal{T}_n^1 = \{y_{ij}^1 = (2^{-n}j, 2^{-n} + 2^{-n/2}i) \colon i = 1, 2, \ldots;\ j = 0, \ldots, 2^n\}.$$

The triangulation with such vertices is represented on Fig. 7.

We set the function φ^n on the vertices from $\mathcal{T}_n^0$ by the equality

$$\varphi^n(y_{ij}^0) = \max\{2^{-n-i}kb \colon 2^{-n-i}kb \leq u_0(2^{-n-i}j\},$$

and on the vertices from $\mathcal{T}_n^1$ by the equality

$$\varphi^n(y_{ij}^1 = \max\{2^{-n}kb \colon 2^{-n-i}kb \leq \varphi(y_{ij}^1)\}.$$

Finally, we set

$$\varphi^n(x, 0) \equiv u_0(x). \tag{66}$$

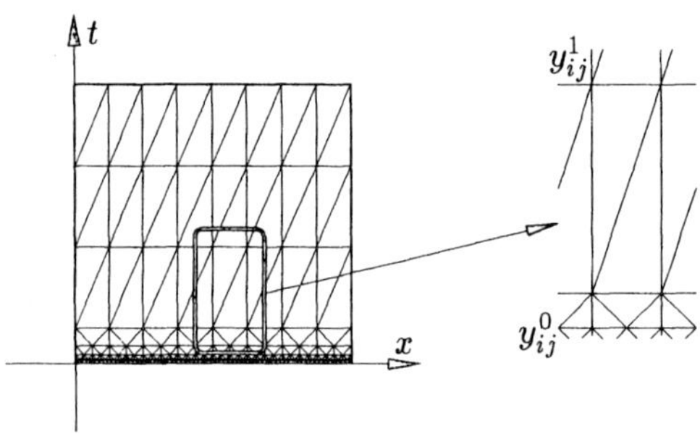

Fig. 7

It is easy to show that $\|\varphi^n - \varphi\|_{C(\mathcal{D})} \longrightarrow 0$ and $\|\varphi_t^n - \varphi_t\|_{L_2(\mathcal{D})} \longrightarrow 0$ as $n \to \infty$, $\|\varphi^n\|_{L_\infty(\mathcal{D})} \leq M < \infty$ for all $n \in \mathbf{N}$, and

$$\varphi_x^n(x, t) \in \{0, 1\} \tag{67}$$

for almost all $(x, t) \in \mathcal{D}$.

Let now $\psi^n = \mathcal{L}^{-1}\varphi_t^n$. By virtue of the results by LADYŽENSKAJA, SOLONNIKOV & URAL'CEVA [?] $\|\psi_t^n\|_{C^\gamma(\mathcal{D})} \leq C \leq \infty$ for some $\gamma > 0$. The latter estimate together with (66) implies smallness of $\psi_x^n(x, t)$ uniformly for small t. But then, because of (67),

$$\psi_t^n = (\psi_x^n)_x - \varphi_t^n = \omega(\varphi_x^n + \psi_x^n)_x - \varphi_t^n$$

(cf. (62)). This means that the function $u^n = \varphi^n + \psi^n$ is a solution of (2). As far as $\|\psi^n - \psi\|_{C(\mathcal{D})} \leq C\|\varphi_t^n - \varphi_t\|_{L_2(\mathcal{D})} \longrightarrow 0$ as $n \to \infty$, it remains to notice that

$$\|u^n - (\varphi + \psi)\|_{C(\mathcal{D})} \longrightarrow \text{ as } n \to \infty.$$

From my point of view a question on correlations between functions from $\mathrm{cl}_{C(\mathcal{D})}(S)$ and measure-valued solutions of the equation (3) described below is rather interesting.

4.4. FBPEs as a junction problem

Let $p_4 < p_5 < p_6 < p_7$ are such that $\omega'(p) \geq d > 0$ for $p \in [p_4, p_5] \cup [p_6, p_7]$, $\omega(p_4) = \omega(p_6)$ and $\omega(p_5) = \omega(p_7)$ (see Fig. 8). In the theorem below existence of a continuum of solutions of Dirichlet BVP for FBPE is asserted. The solutions considered are such that $u_x \in [p_4, p_5]$ on a part of $\mathcal{D}$, and $u_x \in [p_6, p_7]$ on the remaining part, if an initial function u_0 has similar property, namely, if for some $x_0 \in (0,1)$ it satisfies the conditions

$$u_0 \in H_2[0, x_0] \text{ and } u_0'(x) \in (p_4, p_5) \text{ for } x \in [0, x_0],$$

$$u_0 \in H_2[x_0, 1] \text{ and } u_0'(x) \in (p_6, p_7) \text{ for } x \in [x_0, 1],$$

$$\omega[u_0'(x_0 - 0)] = \omega[u_0'(x_0 + 0], \; u_0(0) = u_0(1) = 0.$$

We assume also that $a = b = 0$ in the condition (14).

Theorem (NOVIKOV & SHVAB [34])
For $\|u_0''\|_{L_2}$ and $T > 0$ sufficiently small the problem (2), (13), (14) has a continuum of generalized solutions in $\mathcal{D}$.

This theorem can be proved as follows. The region $\mathcal{D}$ is deviding into two subregions $\mathcal{D}_1$ and $\mathcal{D}_2$ by a segment $\mathcal{I}$ of a line $x = t/k + x_0$ (we suppose that k is sufficiently large, and so this line crosses the boundary of $\mathcal{D}$ in two points $(x_0, 0)$ and $(T/k + x_0, T)$). The problem (2), (13), (14) can be treated as a junction problem, i.e., as problem of finding of functions $v\colon \mathcal{D}_1 \longrightarrow \mathbb{R}$ and $w\colon \mathcal{D}_2 \longrightarrow \mathbb{R}$ such that

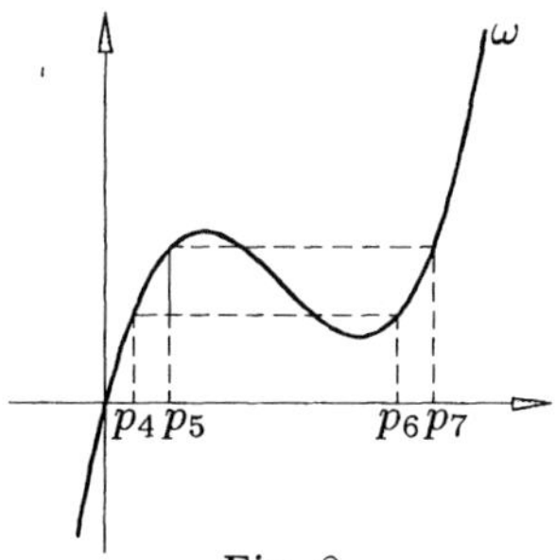

Fig. 8

$$v_t = [\omega(v_x)]_x, \; (x,t) \in \mathcal{D}_1,$$

$$w_t = [\omega(w_x)]_x, \; (x,t) \in \mathcal{D}_2,$$

$$v(x, 0) = u_0(x) \text{ for } x \in [0, x_0], \; w(x, 0) = u_0(x) \text{ for } x \in [x_0, 1],$$

$$v(0,t) = w(1,t) = 0, \; t \in [0, T],$$

$$v = w \text{ and } \omega(v_x) = \omega(w_x) \text{ for } (x,t) \in \mathcal{I}.$$

This problem is regularizing by εu_{xxt}. Classical pass to the limit as $\varepsilon \longrightarrow 0$ proves its solvability in a class of generalized solutions. The proof ends by the inclusions

$$v_x \in [p_4, p_5] \text{ for } (x,t) \in \mathcal{D}_1 \text{ and } w_x \in [p_6, p_7] \text{ for } (x,t) \in \mathcal{D}_2.$$

Arbitrariness in choice of k provides existence of a continuum of solutions.

In NOVIKOV [33] an existence theorem for generalized ($u \in H_1(\mathcal{D}) \cap L_\infty([0,T]), \mathring{W}^2_\infty[0,1])$, $\omega(u_x) \in H_1(\mathcal{D})$) solution of the homogeneous Dirichlet problem for the equation (2) is proven in the case when an initial function $u_0 \in H_2[0,1]$ is such that $u_0'(x)$ lies in a segment entirely containing in parabolicity region $(-\infty, p_1) \cup (p_2, +\infty)$ (see also LAR'KIN, NOVIKOV & YANENKO [25]).

4.5. One more theorem on solvability in special case

Let $c^* \in [c_2, c_1]$ (see Fig. 2), p_1^* and p_2^* are the smallest and greatest roots of the equation $\omega(p) = c^*$. Let us denote the set of functions $u \in W^1_\infty[0,1]$, for which the sets $X_1 = \{x: u'(x) \le p_1^*\}$ and $X_2 = \{x: u'(x) \ge p_2^*\}$ are such that $\mu(X_1 \cup X_2) = 1$, $\mu(X_1 \backslash \operatorname{int} X_1) \cdot \mu(X_2 \backslash \operatorname{int} X_2) = 0$ (μ is the Lebesgue measure on $\mathbb{R}$) by $\mathcal{W}_{c^*}$.

Theorem (PJATKOV [40]). *If $u_0 \in \mathcal{W}_{c^*}$, then for some $T > 0$ BVP (2), (13), (14) has at least one generalized solution such that $u_t \in L_2(\mathcal{D})$, $[\omega(u_x)]_x \in L_2(\mathcal{D})$, $u_x \in L_\infty(\mathcal{D})$ and $u(\cdot, t) \in \mathcal{W}_{c^*}$ for all $t \in [0,T]$.*

The scheme of the proof is as follows. The function ω in the equation (2) is replaced by a nonstrict monotone function

$$\omega^*(p) = \begin{cases} \omega(p) & \text{if } p \in (-\infty, p_1^*] \cup [p_2^*, +\infty), \\ c^* & \text{if } p \in (p_1^*, p_2^*). \end{cases}$$

The resulting BVP (2), (13), (14) received is discretized in time:

$$\frac{u^n - u^{n-1}}{\tau} = [\omega^*(u_x^n)]_x, \; x \in [0,1],$$

$$u^0(x) = u_0(x), \; x \in [0,1], \; u^n(0) = a, \; u^n(1) = b.$$

The solvability of this BVP can be established with the help of strictly monotone approximations of the function ω^*. For approximate solutions $\{u^n\}$ *a priori* estimates (here $N = T/\tau$)

$$\sum_{n=1}^N \tau \left\| \frac{u^n - u^{n-1}}{\tau} \right\|^2_{L_2[0,1]} \le C,$$

$$\sum_{n=1}^N \tau \| [\omega^*(u_x^n)]_x \|^2_{L_2[0,1]} \le C,$$

$$\max_{1 \le n \le N} \| u_x^n \|_{L_\infty[0,1]} \le C$$

are established (cf. with estimates in Subsection 2.4.). *A priori* estimates and the imbedding theorems imply appropriate compactness properties of the family of functions $\{u^n\}_{\tau > 0}$ lifted into $C(\mathcal{D})$. Induction on n proves that $u^n \in \mathcal{W}_{c^*}$ for all n. This fact permits passing to the limit as $\tau \to 0$ in the nonlinearity $\omega^*(u_x^n)$.

We emphasize that in contrast to the well-known result KAMENOMOSTSKAJA (KAMEN) [19], here is proven that $\mu(\{(x,t) \in \mathcal{D}: u_x(x,t) \in (p_1^*, p_2^*)\}) = 0$.

The similar theorem takes place for Neumann BVP (see PJATKOV & PODGAEV [41], PJATKOV [40]).

4.6. Measure-valued solutions of FBPEs

Let us consider degenerate CHE

$$c_t = [\omega(c)]_{xx}, \ (x,t) \in \mathcal{D} \tag{68}$$

(see Subsection 1.2.), which can be reduced to (2) by the change of variables $c = u_x$ and single integration in x. We shall consider Dirichlet problem for it

$$c(0,t) = c(1,t) = 0, \ t \geq 0, \tag{69}$$

$$c(x,0) = c_0(x), \ x \in [0,1]. \tag{70}$$

Just in the same way as in ELLIOTT [92], we define measure-valued solution of the problem (68)–(70). For this purpose, let us remain that a probabilistic measure of a unit weight on $\mathbb{R}$ is called Radon measure (see, f.e., BOURBAKI [9]). We shall denote a value of measure μ on a function $f \in C^1$ by $\langle \mu, f \rangle$, the space of all probabilistic measures on $\mathbb{R}$ by $\mathcal{M}$, and identical mapping on $\mathbb{R}$ by ι: $\iota(x) \equiv x$. A mapping $C: \mathcal{D} \longrightarrow \mathcal{M}$ such that the functions $c(x,t) = \langle C(x,t), \iota \rangle$ and $f(x,t) = \langle C(x,t), \omega \rangle$ satisfy the equality

$$\int_{\mathcal{D}} [c(x,t)\varphi_t(x,t) + f(x,t)\varphi_{xx}(x,t)] \, dx \, dt = 0 \tag{71}$$

for all φ from $C^{2,1}$ with support in $\operatorname{int}\mathcal{D}$ is called *measure-valued solution (solution with values in the space of measures in Tartar-DiPierna sense* (see TARTAR [147], DIPIERNA [90])) of the problem (68)–(70).

We prove existence of weak solution by passing to the limit in difference approximation of the problem (68)–(70).

4.7. Difference solutions and discrete minimum-maximum principle

Let $\mathcal{D}_{\Delta x \Delta t}$ be a uniform grid on $\mathcal{D}$ with steps Δx and Δt, $\Delta x = 1/I$, and $\{c_i^n\}$ be a difference solution of the difference scheme

$$c_i^{n+1} = c_i^n + \tau[\omega(c_{i-1}^n) - 2\omega(c_i^n) + \omega(c_{i+1}^n)], \ 1 \leq i \leq I, \tag{72}$$

$$c_0^{n+1} = c_I^{n+1} = 0, \ c_i^0 = c_0(i\Delta x), \ 0 \leq i \leq I, \tag{73}$$

where $\tau = \Delta t/(\Delta x)^2$, $n = 0,1,2,\ldots$.

Theorem (ELLIOTT [92]) *If ω satisfies the main conditions and Lipschitz condition, then any difference solution satisfies the inequalities*

$$\min\{p_0, c_0^n, \ldots, c_I^n\} \leq e_i^{n+1} \leq \max\{p_0, c_0^n, \ldots, c_I^n\} \tag{74}$$

for all $n = 1,2,\ldots$ and $0 < i < I$ (cf. with Subsection 2.3.).

The proof is a rather standard. By virtue of the Lipschitz condition the function $p \mapsto p - 2\tau\omega(p)$ increases for τ sufficiently small. Let i_0 be such that

$$\omega(e_{i_0}^n) = \min\{\omega(c_0^n), \ldots, \omega(c_I^n)\}. \tag{75}$$

If $c_{i_0}^n = \min\{c_0^n, \ldots, c_I^n\}$, then for $i = 1, \ldots, I-1$

$$c_i^{n+1} = [c_i^n - 2\tau\omega(c_i^n)] + \tau[\omega(c_{i-1}^n) + \omega(c_{i+1}^n)] \geq$$

$$\geq [c_{i_0}^n - 2\tau\omega(c_{i_0}^n)] + \tau[\omega(c_{i_0}^n) + \omega(c_{i_0}^n)] = c_{i_0}^n \geq \min\{p_0, c_0^n, \ldots, c_I^n\}$$

(the cases $i = 0$ and $i = I$ are obvious). If

$$c_{i_0}^n > \min\{c_0^n, \ldots, c_I^n\}, \tag{76}$$

it follows that $c_i^n \geq p_0$ for all $i = 0, \ldots, I$, as far as otherwise (75) and (76) will contradict one to another by virtue of the strict monotonicity of ω to the left of p_0. But then $\omega(c_i^n) \geq \omega(p_0)$, and hence

$$c_i^{n+1} = [c_i^n - 2\tau\omega(c_i^n)] + \tau[\omega(c_{i-1}^n) + \omega(c_{i+1}^n)] \geq$$

$$\geq [p_0 - 2\tau\omega(p_0)] + \tau[\omega(p_0) + \omega(p_0)] = p_0 \geq \min\{p_0, c_0^n, \ldots, c_I^n\},$$

and the left-hand inequality in (74) is proved. The right-hand one be can proved similarly.

4.8. Existence theorem of measure-valued solutions of the problem (68)–(70) (Elliott [92])

If the function ω satisfies the main conditions and the Lipschitz condition, and $c_0 \in C^1$, it follows that (68)–(70) *for any $T > 0$ has at least one measure-valued solution.*

For the proof we construct a sequence of difference solutions $\{c_i^n\}$ of the scheme (72)–(73). Let us denote a suitable piece-wise constant filling of $\{c_i^n\}$ up to a function from $L_\infty(\mathcal{D})$ by $c_\Delta : \mathcal{D} \longrightarrow \mathbb{R}$. Let also $f_\Delta(x,t) \stackrel{\text{def}}{=} \omega[c(x,t)]$. From the discrete minimum-maximum principle it follows that the family $\{c_\Delta\}$, and together with it the family $\{f_\Delta\}$, is bounded in $L_\infty(\mathcal{D})$. Therefore, by virtue of the Alaoglu-Bourbaki theorem, these families are $*$-weakly compact in $L_\infty(\mathcal{D})$. But then (see Tartar [147]) there exist subsequences $\{c_n\} \subset \{c_\Delta\}$ and $\{f_n\} \subset \{f_\Delta\}$, function $c, f \in L_\infty(\mathcal{D})$ and a mapping $C : \mathcal{D} \longrightarrow \mathcal{M}$ such that

$$c_n \to c \text{ and } f_n \to f \ \ *\text{-weakly in } L^\infty(\mathcal{D}),$$

$$c(x,t) = \langle C(x,t), \iota \rangle, \ f(x,t) = \langle C(x,t), \omega \rangle.$$

The identity (71) for any $C^{2,1}$-smooth function with support in $\operatorname{int} \mathcal{D}$ can be checked up in a standard way.

Certainly, for such solutions the question on the equality c and $[\omega(c)]_{xx}$, i.e., on the equality c and f remains open.

Existence theorems in the class of measure-valued solutions for Dirichlet and Neumann BVPs for multidimensional analog of the equation (3) has been proved in Slemrod [143,142]. Proofs are based on passing to the limit in a regularized by $-\varepsilon\Delta^2 u$ equation (for the more details, see Section 5.).

4.9. Asymptotics of measure-valued solutions as $t \longrightarrow \infty$

We shall consider Dirichlet BVP for multidimensional version of (2)

$$u_t = \nabla \cdot \omega(\nabla u), \ (x,t) \in \mathcal{D} \times [0,\infty), \tag{77}$$

$$u(x,t) = 0, \ (x,t) \in \partial\mathcal{D} \times [0,\infty), \tag{78}$$

$$u(x,0) = u_0(x), \ x \in \mathcal{D}. \tag{79}$$

where $\mathcal{D}$ is a bounded domain in $\mathbb{R}^m$, $\omega\colon \mathbb{R}^m \longrightarrow \mathbb{R}^m$ is a continuous potential ($\omega(p) = \nabla\tilde{\omega}(p)$) function, satisfying the following restrictions:

1) $(p, \omega(p)) \geq 0$;
2) $|\omega(p)| \leq C_1(1 + |p|^\gamma)$ $(1 \leq \gamma < 2)$;
3) $C_2|p|^2 - C_3 \leq \tilde{\omega}(p)$;
4) the kernel of the function $p \mapsto (p, \omega(p))$ is bounded.

In SLEMROD [143,142] the BVP derived from (77), (78), (79) by supplying the right-hand part of (77) by regularizing term $-\varepsilon\Delta^2 u$ and completing (78) with the condition $\Delta u(x,t) = 0$, $(x,t) \in \partial\mathcal{D} \times [0,\infty)$, was considered. By passing to the limit as $\varepsilon \longrightarrow 0$, it is proved an existence theorem for global measure-valued solutions of BVP (77), (78), (79). Without monotonicity of the function ω such a solution in general case is not unique.

Let $u \in L_\infty([0,\infty), \overset{\circ}{H}{}^1(\mathcal{D}))$ be one of such solutions issued from sufficiently smooth initial function u_0. We denote an appropriate positive orbit: $O^+(u_0) = \bigcup\limits_{t\geq 0} u(\cdot,t)$ by $O^+(u_0)$ $(\subset \overset{\circ}{H}{}^1(\mathcal{D}))$.

Further, let ρ be a metric of the weak topology on $\overset{\circ}{H}{}^1(\mathcal{D})$, and for any bounded sets $\Omega_1, \Omega_2 \subset \overset{\circ}{H}{}^1(\mathcal{D})$ a *weak distance* between them is defined as $\text{w-dist}(\Omega_1,\Omega_2) = \inf\{\rho(x_1,x_2)\colon x_i \in \Omega_i, \ i = 1,2\}$. *A weak ω-limit set $\omega_w(u_0)$ of an orbit $O^+(u_0)$* is defined in the standard way:

$$\omega_w(u_0) = \{v \in \overset{\circ}{H}{}^1(\mathcal{D})\colon \exists(t_n \longrightarrow +\infty)[u(\cdot,t_n) \overset{(w)}{\longrightarrow} v]\}.$$

It is easy to show that $\omega_w(u_0) \neq \emptyset$ and

$$\text{w-dist}\big(u(\cdot,t), \omega_w(u_0)\big) \longrightarrow 0 \text{ as } t \to \infty.$$

The main result of the work SLEMROD [143,142] is as follows. Weak ω-limit set of orbit of any measure-valued solution of BVP (77), (78), (79) consists of measure-valued stationary solutions of this problem, i.e., for any $v \in \omega_w(u_0)$ there exists a positive probabilistic measure $V(x,t)$ such that

$$\int\limits_{\mathbf{R}^m} \omega(p)\, dV(x,t)(p) = 0 \text{ almost everywhere on } \mathcal{D},$$

and

$$\nabla v(x) = \int\limits_{\mathbf{R}^m} p\, dV(x,t)(p) \text{ almost everywhere on } \mathcal{D}.$$

The similar statements are proved for Neumann BVP for the equation (77).

In KINDERLEHRER & PEDREGAL [120] conditions are specified, for measure-valued solutions coincide with measure-valued solutions of an equation, in which ω is replaced by a convexified in a special manner single-valued function.

In SKOBELEV & YANENKO [44] a linear instability of solutions of the equation (1) is shown at a physical rigidity level.

4.10. Hysteresis solutions of FBPEs

Let us consider a multidimensional analogue of (4) (see Subsection 1.1.)

$$e_t = \Delta\vartheta, \ (x,t) \in \mathcal{D}, \tag{80}$$

where e is an enthalpy, $\vartheta = \omega(e)$ is a temperature of an alloy. The main conditions (see Subsection 2.1.) are assumed to be fulfilled. Besides, we consider that $p_1 = -1$, $p_2 = 1$ (p_1 as used here is a melting point and p_2 is a freezing one), $\omega(p_1) = 1$, $\omega(p_2) = -1$. Let, as usually, f_1, f_2, f_3 are the maximal single-valued branches of ω^{-1}.

The states $(e, f_1(e))$ and $(e, f_3(e))$ meet to metastable phase states of a system, and $(e, f_3(e))$ meets to a hypothetical unstable one.

In PLOTNIKOV [138] the following BVP (*problem B*) for (80) is formulated and investigated assuming absence of unstable states. To find enthalpy e, temperature ϑ and phase concentration λ in $L_\infty(\mathcal{D})$ satisfying the following requirements.

i) For all $(x,t) \in \mathcal{D}$

$$\lambda(x,t) \in [0,1], \ e(x,t) = [1-\lambda(x,t)]f_1[\vartheta(x,t)] + \lambda(x,t)f_3[\vartheta(x,t)];$$

ii) $\lambda(x,t) = 0$ if $\vartheta(x,t) < -1$, and $\lambda(x,t) = 1$ if $\vartheta(x,t) > 1$;

iii) the pair $(e,\vartheta) \in H_1(\mathcal{D}) \times H_1(\mathcal{D})$ is a generalized solution to BVP for (80) separated out by the boundary and initial conditions

$$\frac{\partial\vartheta}{\partial n} = 0, \ (x,t) \in \partial\mathcal{D} \times [0,T],$$

$$e(x,0) = e_0(x), \ \vartheta(x,0) = \vartheta_0(x), \ x \in \mathcal{D};$$ [3]

iv) $\lambda(x,0) = \lambda_0(x)$ for $x \in \mathcal{D}$;

v) For any non-decreasing C^2-smooth function φ

$$\frac{\partial}{\partial t}\left(\Phi_1(x,t)[1-\lambda(x,t)] + \Phi_3(x,t)\lambda(x,t)\right) \leq \varphi[\vartheta(x,t)]e_t(x,t), \ (x,t) \in \mathcal{D},$$

[3] I.e. for any smooth function φ on $\mathcal{D}$ equal zero for $t = T$

$$\int_{\mathcal{D}} (\nabla\vartheta\cdot\nabla\varphi - e\varphi_t)\,dx\,dt = \int_{\mathcal{D}} e_0\varphi(x,0)\,dx,$$

$$\|\vartheta(\cdot,t)-\vartheta_0\|_{H_0(\mathcal{D})} + \|e(\cdot,t)-e_0\|_{H_0(\mathcal{D})} \to 0 \text{ as } t\to 0.$$

where

$$\Phi_1(x,t) = \int_{-1}^{\vartheta(x,t)} \varphi(s) f_1'(s)\, ds,$$

$$\Phi_3(x,t) = \int_{-1}^{\vartheta(x,t)} \varphi(s) f_3'(s)\, ds + \int_{-1}^{1} \varphi(s)[f_1'(s) - f_3'(s)]\, ds.$$

The condition v) in some sense is a generalized Clausius-Duhem law.

Below we suppose that $\lambda_0 \in L_\infty(\mathcal{D})$, $\lambda_0(x) \in [0,1]$ for $x \in \mathcal{D}$, $\vartheta_0 \in L_\infty(\mathcal{D}) \cap H_1(\mathcal{D})$, $||\vartheta_0||_{C^*(\mathcal{D})} < \infty$, $e_0 \in f_1(\vartheta_0(x)) + [f_3(\vartheta_0(x)) - f_1(\vartheta_0(x))] \times \Gamma(\lambda_0(x), \vartheta_0(x))$, where $\Gamma\colon [0,1] \times \mathbb{R} \to 2^{\mathbf{R}}$ is a multi-valued mapping set by the formula

$$\Gamma(x,y) = \begin{cases} 0 & \text{for} \quad (x,y) \in [0,1] \times (-\infty,-1), \\ [0,x] & \text{for} \quad (x,y) \in [0,1] \times \{-1\}, \\ x & \text{for} \quad (x,y) \in [0,1] \times (-1,-1), \\ [x,1] & \text{for} \quad (x,y) \in [0,1] \times \{1\}, \\ 1 & \text{for} \quad (x,y) \in [0,1] \times (1,\infty). \end{cases}$$

Theorem. *The problem B has a generalized solution such that*

$$||\vartheta||_{H_1(\mathcal{D})} + ||\Delta\vartheta||_{L_\infty([0,T],C^*(D))} \le C,$$

$$||e(\cdot,t_1) - e(\cdot,t_2)||_{L_1(\mathcal{D})} \le C|t_1 - t_2|,\ t_1, t_2 \in [0,T].$$

The theorem can be proved by a time discretization of the problem and subsequent passage to the limit (cf. with Subsection 6.1.).

We shall note that the family of probabilistic measures

$$V(x,t) = [1 - \lambda(x,t)]\delta\big(\cdot - f_1[\vartheta(x,t)]\big) + \lambda(x,t)\delta\big(\cdot - f_3[\vartheta(x,t)]\big)$$

is a measure-valued solution of FBPE

$$e_t = \Delta\omega(e).$$

Let us define on $\mathcal{D}$ the Radon measure $\mathcal{S}$ by the equality

$$\langle \mathcal{S}, \zeta \rangle = -\int_{\mathcal{D}} \big([h_1(x,t) - h_3(x,t)]\zeta(x,t)\big)_t \lambda(x,t)\, dx\, dt,\ \zeta \in \mathring{C}_\infty(\mathcal{D}),$$

where

$$h_i(x,t) = \Phi_i(x,t) - \varphi[\vartheta(x,t)] f_i[\vartheta(x,t)].$$

The following theorem shows a hysteresis character of the solutions of the problem B.

Theorem. *For any solution (e, ϑ, λ) and any non-decreasing C^2-smooth function φ on $\mathbb{R}$ the following statements take place.*

1) The Radon measure $\mathcal{S}$ can be represented in the form

$$\langle \mathcal{S}, \zeta \rangle = \int\limits_{\mathcal{D}} \langle \sigma_x, \zeta(x, \cdot) \rangle \, dx,$$

where a Radon measure σ_x on $(0, T)$ is such that

$$\|\sigma_x\| \leq C(1 + \|\vartheta_t(x, \cdot)\|_{L_1[0,T]})$$

(C is a constant depending on φ and $(e_0, \vartheta_0, \lambda_0)$).

2) For almost all $x \in \mathcal{D}$ the function $\lambda(x, \cdot)$ is of bounded variation on $(0, T)$. Besides, if $\vartheta(x, t) < 1$ (respectively, $\vartheta(x, t) > -1$) on some interval (t_1, t_2), then $\lambda(x, \cdot)$ is nonincreasing (respectively, nondecreasing) on this interval.

3) The function $\mu(x, t) = \lambda_t(x, t)$ is an absolutely continuous charge with respect to the Lebesgue measure:

$$\langle \mu, \zeta \rangle = \int\limits_{\mathcal{D}} \langle \mu_x, \zeta \rangle \, dx$$

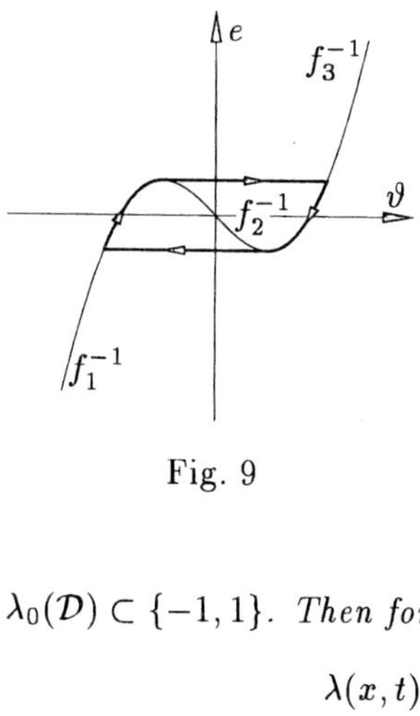

Fig. 9

Statement 2) of the theorem shows that the hysteresis loop goes around in the negative direction (see Fig. 9).

In the following theorem it is asserted that in thermodynamically isolated systems the pure phases do not pass to a metastable mixture.

Theorem. *Let ω is piece-wise linear function:*

$$\omega(p) = \begin{cases} p + 2 & \text{for} \quad p \leq -1, \\ -p & \text{for} \quad -1 < p < 1, \\ p - 2 & \text{for} \quad p \geq 1, \end{cases}$$

and the initial data of the problem do not lie inside the hysteresis loop: $\vartheta_0(\mathcal{D}) \cap \{-1, 1\} = \emptyset$ and $\lambda_0(\mathcal{D}) \subset \{-1, 1\}$. Then for almost all $(x, t) \in \mathcal{D}$

$$\lambda(x, t) \in \{0, 1\} \text{ and } \vartheta(x, t) = \omega[e(x, t)].$$

4.11. Asymptotic behavior solutions of the model of non-Newtonian fluid

The BVP

$$u_t = \sigma_x + u_{xx} + f, \ (x, t) \in [0, 1] \times [0, \infty), \tag{81}$$

$$\sigma_t + \sigma = g(u_x), \ (x, t) \in [0, 1] \times [0, \infty), \tag{82}$$

$$\sigma(0, t) + u_x(0, t) = 0, \ u(1, t) = 0, \ t \in [0, \infty), \tag{83}$$

$$u(x,0) = u_0(x),\ \sigma(x,0) = \sigma_0(x),\ x \in [0,1] \tag{84}$$

describes (see Subsection 1.6.) shear flows of a non-Newtonian fluid between two plates $x = \pm 1$ (a flow is assumed to be symmetric about the plane $x = 0$). If the function $\omega\colon p \mapsto g(p) + p$ is not monotone, it follows that stationary solutions of this system satisfying the equation

$$[\omega(u_x)]_x = 0,$$

are "splicing" of the solutions of the equations

$$u_x = g_i(fx),$$

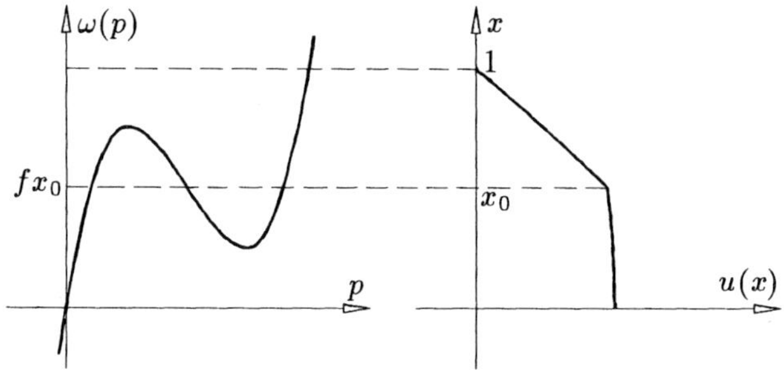

Fig. 10

where g_i are the maximal continuous single-valued branches of ω^{-1}. Piece-wise smooth solutions are profiles with kinks (see Fig. 10). It turn out, that solutions (81)–(84) issued from sufficiently smooth initial data close to a stationary solution outside of a neighborhood of kink points tends to the given stationary solution as $t \longrightarrow \infty$, if the stationary solution lies on segments of growth of ω. More exactly, let (u^*, σ^*) be a stationary solution of the problem (81)–(84) such that u^* has a finite number of kinks $x_1, \dots, x_n$.

Theorem (Nohel, Pego & Tzavaras [131]). *Let*

$$\omega'[u_x^*(x)] \geq p_0 > 0,\ x \in [0,1],\ x \neq x_1, \dots, x_n,$$

and $\mathcal{U}$ be a neighborhood of the set $\{x_1, \dots, x_n\}$. Then for mes$\mathcal{U}$ sufficiently small there exists $\delta_0 > 0$ such that for $0 < \delta < \delta_0$ and for any smooth initial data (u_0, σ_0) such that

$$\sup_{0 \leq x \leq 1} |\sigma_0(x) + v_{0x}(x) + fx| < \delta,$$

$$\int_0^1 u_t(x,0)\, dx < \frac{1}{2}\delta^2,$$

$$|U_{0x}(x) - u_x^*(x)| < \delta,\ x \in [0,1] \backslash \mathcal{U},$$

appropriate solution (u, σ) stabilizes to (u^*, σ^*):

$$u_x(x,t) \longrightarrow u^*(x),\ \sigma(x,t) \longrightarrow \sigma^*(x)$$

as $t \longrightarrow \infty$ for all $x \in [0,1] \backslash \mathcal{U}$.

4.12. On a structure of separation boundary of forward and backward parabolicity regions

GREBENEV in [108] considers the following filtration FBPE

$$u_t = \frac{1}{2}(u^2)_{xx}, \ t \in \mathbb{R} \times (0,T) \ (= \mathcal{D}). \tag{85}$$

This equation can be rewritten in a non-divergent form:

$$u_t = uu_{xx} + u_x^2.$$

If u is of variable sign, then (85) is FBPE. By *a continuous weak solution* of the equation (85) we mean a function $u \in C(\mathcal{D}) \cap L^\infty(\mathcal{D})$ such that $u_x \in L^\infty_{\text{loc}}(\mathcal{D})$, and for any $0 < t_1 < t_2 < T$, $x_1 < x_2$ and function $\varphi \in C^{1,1}(\mathcal{D})$ the identity

$$\int_{x_1}^{x_2} u(x,t_2)\varphi(x,t_2)\,dx - \int_{x_1}^{x_2} u(x,t_1)\varphi(x,t_1)\,dx = \int_{t_1}^{t_2}\int_{x_1}^{x_2} (u\varphi_t - uu_x\varphi_x)\,dx\,dt$$

is fulfilled.

The set

$$\mathcal{J} = \{(x,t) \in \mathcal{D} : u(x,t) = 0\}$$

is said to be *an interface* of the solution, the set

$$\mathcal{D}^+ = \{(x,t) \in \mathcal{D} : u(x,t) > 0\}$$

is said to be *a forward parabolicity region*, and the set

$$\mathcal{D}^- = \{(x,t) \in \mathcal{D} : u(x,t) < 0\}$$

is said to be *a backward parabolicity region.* It is easy to show that in $\mathcal{D}\backslash\mathcal{J}$ a continuous weak solution is classical one.

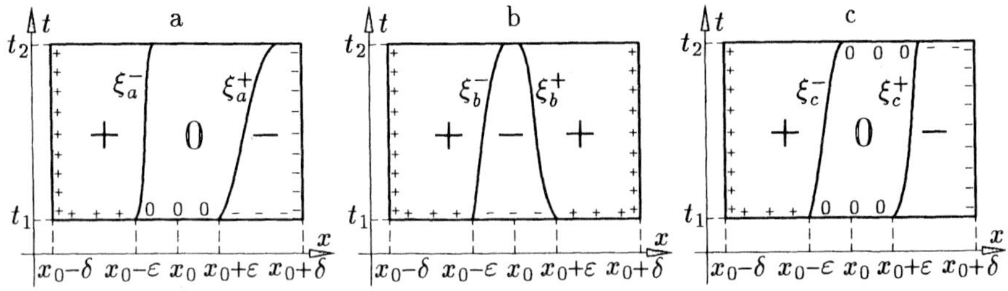

Fig. 11

A local behavior of interfaces boundaries is investigated in GREBENEV [108]. Denote the rectangular $[x_0-\delta, x_0+\delta] \times [t_1,t_2]$ by R, and let u be a continuous weak solution of the equation (85) in R. Then:

i) If $u(x,t) > 0$ for $(x,t) \in (\{x_0-\delta\} \times [t_1,t_2]) \cup ([x_0-\delta, x_0-\varepsilon] \times \{t_1\})$ (where $0 \le \varepsilon < \delta$) and $u(x,t) < 0$ for $(x,t) \in (\{x_0+\delta\} \times [t_1,t_2]) \cup ([x_0+\varepsilon, x_0+\delta] \times \{t_2\})$, then boundaries of the interface in R are continuous curve parametrized by the time t. More exactly, there exist continuous functions $\xi_a^+, \xi_a^-: [t_1,t_2] \longrightarrow [x_0-\delta, x_0+\delta]$ such that

$$\mathcal{J} \cap R = \{(x,t) \in R: t \in [t_1,t_2],\ x \in [\xi_a^-(t), \xi_a^+(t)]\}$$

(see Fig. 11a).

ii) If $u(x,t) > 0$ for $(x,t) \in (\{x_0-\delta\} \times [t_1,t_2]) \cup ([x_0-\delta, x_0-\varepsilon] \times \{t_1\}) \cup (\{x_0+\delta\} \times [t_1,t_2]) \cup (x_0+\varepsilon, x_0+\delta) \times \{t_2\})$ (where $0 < \varepsilon < \delta$) and $u(x,t) < 0$ for $(x,t) \in (\{x_0-\varepsilon, x_0+\varepsilon) \times \{t_1\}$, then the interface in R is a union of two continuous curve, parametrized by the time t. More precisely, there exist continuous functions $\xi_b^+, \xi_b^-: [t_1,t_2] \longrightarrow [x_0-\delta, x_0+\delta]$ such that

$$\mathcal{J} \cap R = \{(x,t) \in R: t \in [t_1,t_2],\ x \in \{\xi_b^-(t)\} \cup \{\xi_b^+(t)\}\}.$$

In particular, this means that a backward parabolicity region can not be destroyed inside R (see Fig. 11b).

iii) If $u(x,t) > 0$ for $(x,t) \in (\{x_0-\delta\} \times [t_1,t_2]) \cup ([x_0-\delta, x_0-\varepsilon] \times \{t_1\})$ (where $0 \le \varepsilon < \delta$) and $u(x,t) < 0$ for $(x,t) \in (\{x_0+\delta\} \times [t_1,t_2]) \cup (x_0+\varepsilon, x_0+\delta] \times \{t_2\})$, then the boundary of interface in R is a union of two continuous monotone curve parametrized by the time t. More explicitly, there exist a continuous non-decreasing functions $\xi_c^+, \xi_c^-: [t_1,t_2] \longrightarrow [x_0-\delta, x_0+\delta]$ such that

$$\mathcal{J} \cap R = \{(x,t) \in R: t \in [t_1,t_2], \quad x \in \{\xi_c^-(t)\} \cup \{\xi_c^+(t)\}\}$$

(see Fig. 11c).

An information on local behavior of interfaces allows (see GREBENEV [108]) to prove a local solvability of the Cauchy problem for the equation (85) with initial condition

$$u(x,0) = u_0(x),\ x \in \mathbb{R},$$

in the class of continuous weak solutions, when the function u_0 vanishes on some segment $[x_1,x_2]$ $(x_1 \le x_2)$ and is such that $u_0(x) > 0$ for $x < x_1$ and $u_0(x) < 0$ for $x > x_2$.

Solvability of some non-classical BVPs for the equation (85) and some its extensions with boundary conditions

$$u(x,0) = u_1(x),\ x \le 0;$$

$$u(x,T) = u_2(x),\ x \ge 0,$$

in the class of continuous weak solutions, when $u_1(x) > 0$, $u_2(x) < 0$ under additional growth restrictions for u_1 and u_2 at zero is investigated in GREBENEV [11].

4.13. Numerical investigations of FBPEs

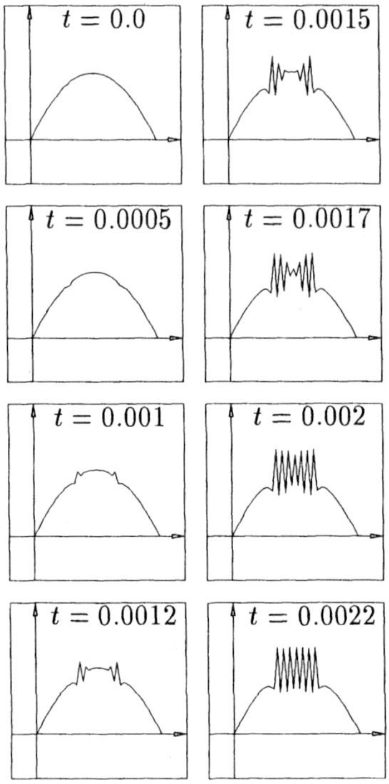

Fig. 12

Complexity of FBPEs as mathematical object increases an interest to their numerical studies. Only preliminary results are received in this direction. Naturally, during numerical modeling of such equations difficulties concerning with ill-posedness BVPs come up.

We have restricted ourselves to listing some of works (far from all) completely or partially devoted to numerical analysis of such equations: ELLIOTT [92], HÖLLIG & NOHEL [118], KIM & ZELINSKAJA [20], MASLOVA [27], MASLOVA, NOVIKOV & YANENKO [28], STRANG & ABDEL-NABY [146], VORONKO [10].

By way of illustration we give the simplest example of initial process of evolution of a numerical solution of BVP (68)–(70) in the case of piece-wise linear function ω and square-law initial function on $[0, 1]$. This solution is obtained through the difference scheme (72)–(73) ($\Delta x = 0.025$, $\Delta t = 5 \cdot 10^{-6}$). It demonstrates a strong instability and oscillatory character. Certainly, in the computations a scheme effects play a strong role.

5. Regularizations of forward backward parabolic equations per se

FBPEs can be regularized in various ways. In this Section we give (of necessity brief) the review of results concerning with equations regularizing FBPEs as to independent objects of discussion. Some of them will be described namely as regularizators in the next Section.

5.1. Regularization of the equation (1) by the operator $\varepsilon \dfrac{\partial^5}{\partial x^4 \partial t}$

The BVP

$$\left(1 + \varepsilon \frac{\partial^5}{\partial^4 x \partial t}\right) u_t + u u_x = [\omega(u_x)]_x, \ (x, t) \in \mathcal{D}, \tag{86}$$

$$u(0,t) = u(1,t) = u_x(0,t) = u_x(1,t) = 0, \ t \in [0, T], \tag{87}$$

$$u(x, 0) = u_0(x), \ x \in [0, 1] \tag{88}$$

can be considered as a regularization of the homogeneous Dirichlet BVP (1), (13), (14). It is natural to call a function $u \in L_\infty([0,T], \overset{\circ}{H}{}^2[0,1])$ such that $u_t \in L_2([0,T], \overset{\circ}{H}{}^2[0,1])$ and

$$(u_t, v) + \varepsilon \frac{\partial}{\partial t}(u_{xx}, v_{xx}) + (\omega(u_x), v_x) + (uu_x, v) = 0$$

for all $v \in \overset{\circ}{H}{}^2[0,1]$, where $(\cdot,\cdot)$ is a scalar product in $L_2[0,1]$, a *generalized solution* of the problem (86)–(88).

Theorem (ZELENJAK, NOVIKOV & YANENKO [?]). *If ω is a cubic polynomial satisfying the main conditions, and $u_0 \in \overset{\circ}{H}{}^2[0,1]$, it follows that* (86)–(88) *has at least one generalized solution in $\mathcal{D}$ for some $T > 0$.*

The proof of this theorem is a rather standard (cf. LIONS [26]). Let e_i be an orthonormal in $L_2[0,1]$ system of eigenfunctions for the operator $-\dfrac{d^2}{dx^2}$ with the homogeneous Dirichlet boundary conditions. Coefficients α^{nk} of Galerkin approximation $u^n(x,t) = \sum_{k=1}^n \alpha^{nk}(t)e^k(x)$ are determined from the system of ordinary differential equations

$$(u_t^n, e^l) + \varepsilon(u_{txx}^n, e_{xx}^l) + (\omega(u_x^n), e_x^l) + (u^n u_x^n, e^l) = 0, \ l = 1, \dots, n, \tag{89}$$

and initial conditions

$$\sum_{k=1}^n \alpha^{nk}(0)e^k = u_0. \tag{90}$$

Multiplication (89) by α^{nl} and summation in l easily proves the *a priori* estimate

$$||u_t^n(\cdot,t)||_{L_2[0,1]} + \varepsilon ||u_{xx}^n(\cdot,t)||_{L_2[0,1]} \le C, \tag{91}$$

which guarantees, by virtue of the Peano theorem, a solvability of the problem (89)–(90) on $[0,T]$. Multiplication (89) by $(\alpha^{nl})'$, summation in l and integration in t from zero to τ lead to the *a priori* estimate

$$||u_t^n||_{L_2(\mathcal{D}_\tau)} + \varepsilon ||u_{txx}^n||_{L_2\mathcal{D}_\tau)} + ||u_x(\cdot,\tau)||_{L_4[0,1]} \le C \tag{92}$$

(here we used the evident estimate $\int_0^p \omega(s)\,ds \ge ap^4 + b$). The latter implies $*$-weak compactness of $\{u^n\}$ in $L_\infty([0,T], \overset{\circ}{H}{}^2[0,1])$, weak compactness of $\{u_t^n\}$ in $L_2([0,T], \overset{\circ}{H}{}^2[0,1])$ and, as far as ω is a cubic polynomial, weak compactness of $\{\omega(u_x)\}$ in $L_\infty([0,T], L_{4/3}[0,1])$. Therefore, without loss of generality, there exist $u \in L_\infty([0,T], \overset{\circ}{H}{}^2[0,1])$ such that $u_t \in L_\infty([0,T], \overset{\circ}{H}{}^2[0,1])$ and $v \in L_\infty([0,T], L_{4/3}[0,1])$ such that

$$u^n \longrightarrow u \ *\text{-weakly in } L_\infty([0,T], \overset{\circ}{H}{}^2[0,1]),$$

$$u_t^n \longrightarrow u_t \text{ weakly in } L_2([0,T], \overset{\circ}{H}{}^2[0,1]), \tag{93}$$

$$\omega u_x^n \longrightarrow v \text{ weakly in } L_{4/3}(\mathcal{D}),$$

as $n \to \infty$. Estimates (91) and (92) imply a uniform with respect to n boundedness of $\{u_x^n\}$ in $\overset{\circ}{H}{}^1[0,1](\mathcal{D})$. This allows to consider that $u_x^n \longrightarrow u_x$ strongly in $L_2(\mathcal{D})$ and almost everywhere on $\mathcal{D}$. But then, in view of Lemma 1.3 from LIONS [26], (93) leads to the equality $\omega(u_x) = v$. Besides, $u^n \longrightarrow u$ strongly in $L_2(\mathcal{D})$. Passing to the limit in (89) completes the proof.

5.2. The Cahn-Hilliard equation

As already indicated CHE

$$c_t + \varepsilon c_{xxxx} = [\omega(c)]_{xx} \tag{94}$$

fulfills the role of natural regularization for the equation (3). The regularizing term εc_{xxxx} corresponds to energy of surface (capillary) interaction caused by a space non-homogeneity of concentration. This equation attracts active interest of researchers initiated by the pioneer work by CAHN & HILLIARD [83]. We shall describe here only some results relating (94) which are important, in our opinion, for discussing the equation (3). The review of the recent work on the CHE can be found in ELLIOTT [99].

We shall assume (unless otherwise specified) that ω is a cubic polynomial with positive at infinity derivative possessing three real roots, one of which is zero: $\omega(p) = \gamma_3 p^3 + \gamma_2 p^2 + \gamma_1 p$ $(\gamma_2^2 > 4\gamma_1\gamma_2,\ \gamma_3 > 0)$.

In this subsection we give (of necessity brief) review of results concerning CHE as independent object of investigation.

We consider a multidimensional variant of CHE (94) in a bounded domain $\mathcal{D} \in \mathbb{R}^n$ $(n \leq 3)$ with smooth boundary Γ

$$c_t + \varepsilon \Delta^2 c = \Delta\omega(c),\ \ (t,x) \in \mathcal{D}; \tag{95}$$

where $\mathcal{D} = \mathcal{D} \times [0,T]$, Δ is the Laplas operator. Let us consider for this equation the BVP

$$\left.\frac{\partial c}{\partial n}\right|_{\Gamma} = 0,\ \ \left.\frac{\partial \Delta c}{\partial n}\right|_{\Gamma} = 0, \tag{96}$$

$$c(x,0) = c_0(x),\ \ x \in \mathcal{D}, \tag{97}$$

where n is an external normal to the boundary Γ, and c_0 is a given function.

The no-flux conditions (96) obviously guarantee mass conservation law

$$\int_{\mathcal{D}} c(x,t)\,dx \equiv m \cdot \mathrm{mes}\mathcal{D},$$

where $m \overset{\text{def}}{=} \int_{\mathcal{D}} c_0(x)\,dx$.

If $c_0 \in H^1(\mathcal{D})$, then the global existence and uniqueness theorem of generalized solutions c such that $c \in L^\infty([0,T], H^1(\mathcal{D}))$, $\omega(c) - \Delta c \in L^2([0,T], H^1(\mathcal{D}))$, $c_t \in L^2([0,T], (H^1(\mathcal{D}))')$ takes place for the problem (95)–(97). The *a priori* estimates needed for proof of this theorem follow from the fact that energy functional

$$\mathcal{F}_\varepsilon(c) = \int\limits_{\mathcal{D}} \left[W(c) + \frac{\varepsilon}{2}|\nabla c|^2\right] dx,$$

$\left(W = \int\limits_{\mathcal{D}} \omega\, dx\right)$ is a Lyapunov functional. Namely,

$$\frac{d\mathcal{F}_\varepsilon(c(\cdot,t))}{dt} = -\|\nabla[\omega(c(\cdot,t)) - \varepsilon\Delta c(\cdot,t)]\|_{L^2(\mathcal{D})} \leq 0 \tag{98}$$

on solutions.

If $c_0 \in \overset{\circ}{H}{}^2(\mathcal{D}) \overset{\text{def}}{=} \left\{c \colon c \in H^2(\mathcal{D}),\ \frac{\partial c}{\partial n}\big|_\Gamma = 0\right\}$, then $c \in L_2([0,T], H^4(\mathcal{D}))$, $c_t \in L_2([0,T], L_2(\mathcal{D}))$. If $c_0 \in H^6(\mathcal{D}) \cap \overset{\circ}{H}{}^2(\mathcal{D})$ and $\Delta c_0 \in \overset{\circ}{H}{}^2(\mathcal{D})$, then a solution c is a classical one.

These statements, their extensions, existence theorems for Dirichlet BVP defined by the conditions

$$c|_\Gamma = c_1(x),\ [\omega(c) - \varepsilon\Delta c]|_\Gamma = c_2(x),$$

and also for periodic BVP on the torus $\mathcal{D} = [0, X]^n$ have been proved in ELLIOTT & ZHENG [98], ZHENG [152], ELLIOTT [99], ELLIOTT & LARSSON [95], NICOLAENKO, SCHEURER & TEMAM [130] and others.

If γ_3 in polynomial representation of the function ω is negative, then solutions of the scalar problem (95)–(97) blows up in a finite time, i.e., for some $T < \infty$

$$\lim_{t \to T} \|c(\cdot,t)\|_{H^2(\mathcal{D})} = +\infty$$

(see ELLIOTT & ZHENG [98]).

Existence theorems for BVP (95)–(97) type for equations of CHE type, derived from non-differentiable energy functionals $\mathcal{F}_0(c) + \alpha \int\limits_{\mathcal{D}} [\beta^2 + |\nabla c|^2]^{1/2}\, dx$, $\mathcal{F}_0(c) + \alpha \int\limits_{\mathcal{D}} |\nabla c|\, dx$ and $\int\limits_{\mathcal{D}} [W(c) + \alpha|\nabla c|]\, dx$ have been proved in ELLIOTT & MIKELIĆ [97], for an equation taking into account a mobility a mixture components in YIN [150], for equations describing a non-isothermic separation of mixtures in ALT & PAWLOW [67,68].

If a trajectory of solution of the scalar problem (95)–(97) starts in $\overset{\circ}{H}{}^4(\mathcal{D}) \overset{\text{def}}{=} \left\{c \colon c \in \overset{\circ}{H}{}^2(\mathcal{D}) \cap H^4(\mathcal{D}) \colon \frac{\partial \Delta c}{\partial n}\bigg|_\Gamma = 0\right\}$, then it is pre-compact in $H^2(\mathcal{D})$ ZHENG [152]. Therefore, ω-limit set $\Omega(c_0)$ of such a trajectory is compact and connected in $H^2(\mathcal{D})$. Besides, because of $\mathcal{F}_\varepsilon$ is a Lyapunov functional, it is a constant on $\Omega(c_0)$. From (98) directly follows (ZHENG [152]) that any function from $\Omega(c_0)$ is a solution of the stationary problem. The latter after the double integration with respect to x can be rewritten in the form

$$\varepsilon\Delta c = \omega(c) - \sigma, \ x \in \mathcal{D}, \tag{99}$$

$$\left.\frac{\partial c}{\partial n}\right|_{\Gamma} = 0, \tag{100}$$

$$\int\limits_{\mathcal{D}} c(x)\,dx = m; \tag{101}$$

where σ is a constant to be defined.

A representation of a general solution for the scalar equation (99) in terms of elliptic Jacobi functions see in NOVICK-COHEN & SEGEL [132].

On the set of classical solutions of scalar BVP (99)–(101) is equivalent (ZHENG [152]) to a problem on critical points of energy functional $\mathcal{F}_\varepsilon$ on $\{c \in H^1(\mathcal{D}){:}\ \int_{\mathcal{D}} c\,dx = m\}$. If ε is sufficiently large, then this problem has no non-trivial solutions, and any solution of the problem (96)–(97) started in $\overset{\circ}{H}{}^4(\mathcal{D})$ converges to the trivial solution in $H^2(\mathcal{D})$. The latter statement holds also if (for any fixed ε) $|m|$ is sufficiently large.

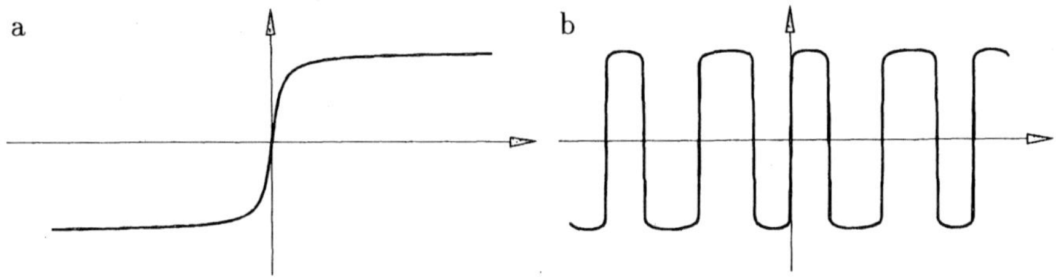

Fig. 13

In CARR, GURTIN & SLEMROD [84] existence of solutions for the one-dimensional problem (99)–(101) is proved (the domain $\mathcal{D}$ is assumed to be a finite interval). It is shown that for small ε this problem has two monotone solutions (see Fig. 13a; the second monotone solution is symmetrical to the depicted on the Fig. 13a one with respect to the center of interval) and "periodic" in x solutions (see Fig. 13b). It is proved also that only monotone solutions provide a minimum to energy functional, and periodic solutions are its saddle points (for more details see Subsection 6.3.). ALEXIADES & AIFANTIS [59] discuss similar questions in the case of unbounded interval of definition of solution. Moreover, in one-dimensional case it is proved (ZHENG [152]) that for small ε the problem (99)–(101) has exactly $2[C/\varepsilon] + 1$ solutions, where C is a constant effectively calculated via data of the problem, and $[\cdot]$ denotes an integral part of a number. Any solution of the problem (95)–(97) started in $\overset{\circ}{H}{}^4(\mathcal{D})$ converges to one of these stationary solutions in $H^2(\mathcal{D})$.

There is a series of works devoted to numerical analysis of CHE: BROWLEY & ELLIOTT [80], COPETTI & ELLIOTT [88], ELLIOTT [99], ELLIOTT &

FRENCH [93,94], ELLIOTT, FRENCH & MILNER [100], ELLIOTT & LARSSON [95], FURIHATA, ONDA & MORI [104,105] e.a.

5.3. Asymptotic behavior of the Cahn-Hilliard equation as $t \to \infty$

Since the energy functional $\mathcal{F}_\varepsilon$ is a Lyapunov functional, it follows that limiting as $t \longrightarrow \infty$ solutions should be its local minima on the subspace selected by mass conservation condition (see HENRY [48], ALIKAKOS, BATES & FUSCO [65]).

In BATES & FIFE [73] a stability properties of monotone equilibria of the one-dimensional CHE are discussed. A metastable constant solutions and solution with boundary and internal layers are considered.

Local stability of non-monotone equilibria with respect to special (spatially located) perturbations is shown in NAUMAN & BALSARA [129].

In RANKIN, III [139] it is shown that the CHE can be rewritten as a half-linear differential equation in a Banach space with a sectorial in the sense of HENRY [48] operator, and hence a linear part of the equation generates an analytical semigroup.

A nonlinear semigroup $\mathcal{T}(t)$ generated by BVP (95)–(97) is a C^1-smooth semigroup of completely continuous operators on $\overset{0}{H}{}^1(\mathcal{D})$ (ELLIOTT & LARSSON [95]). Presence of Lyapunov functional for (95) implies existence of an attractor $\mathring{A}$ of the semigroup $\mathcal{T}(t)$ (ELLIOTT & LARSSON [95]).

Local structure of this attractor in one-dimensional case was investigated in ALIKAKOS, BATES & FUSCO [65], BATES & XUN [75,76]. Let us call a solution of the stationary one-dimensional problem *N-layered*, if it contains N parts of fast variation (thus, (see previous subsection) among N-layered solutions only 1-layered one are stable). Behavior of $\mathcal{T}(t)$ in a neighborhood of contained in $\mathring{A}$ one-dimensional invariant set connecting 1-layered and 2-layered stationary solutions (entering separatrix of 2-layered saddle point) is investigated in ALIKAKOS, BATES & FUSCO [65]. Roughly speaking, it is proved that along this manifold a solution moves with a velocity $O(e^{-C/\varepsilon})$, and in a normal direction this variety attracts solutions with a velocity uniformly in ε separated from zero. Topological structure of a neighborhood of $(N+1)$-layered equilibria is investigated in BATES & XUN [75,76] (see also BRONSARD & HILHORST [79]). In particular, it is shown that they are hyperbolic points with N-dimensional unstable manifold. Behavior of $\mathcal{T}(t)$ in a neighborhood of this manifolds is described .

GRANT [107] investigates a behavior of solutions of Neumann BVP for the one-dimensional CHE (94) with boundary and initial conditions

$$c_x(0,t) = c_x(1,t) = c_{xxx}(0,t) = c_{xxx}(1,t) = 0, \ t > 0, \tag{102}$$

$$c(x,0) = c_0(x), \ x \in [0,1] \tag{103}$$

in a neighborhood of above periodic equilibria in the case when the mass $m = \int_0^1 c_0(x)\,dx$ of component A lies in a spinodal interval, i.e., $\omega'(m) < 0$. It is easy to see that linearized in a neighborhood of equilibrium $c(x) \equiv m$ equation has a one-dimensional subspace $\{\alpha \cdot \cos n_0\pi x : \alpha \in \mathbb{R}\}$ of fastest growing solutions. Let Ω be a neighborhood in $H^2(0,1) \cap \{c: \int_0^1 c(x)\,dx = m,\ c'(0) = c'(1) = c'''(0) = c'''(1) = 0\}$ of the set of all $(2/kn_0)$-periodic equilibria of the equation (94), and $A(\Omega)$ be a set of initial functions such that the corresponding solution of (94), (102), (103) enters Ω in a finite time.

Theorem (GRANT [107]). *If a function c_0 is not H^2-orthogonal to the function $x \mapsto \cos n_0\pi x$, then $m + \rho(c_0 - m) \in \Omega$ for small ρ.*

Define the projection P by the equality $P\big(m + \sum_{n=0}^{\infty} a_n \cos n\pi x\big) = m + a_{n_0} \cos n_0\pi x$. Let $\lambda_{n_0} > \lambda_{n_1}$ are two greatest eigenvalues of the linearized along $c(x) \equiv m$ problem.

Theorem (GRANT [107]). *If $p < \min\{\lambda_{n_0}/\lambda_{n_1}, 2\}$, it follows that there exist such $C, r > 0$ that the inequalities*

$$C\|c_0 - Pc_0\|_{H^2}^p < \|Pc_0 - m\|_{H^2} < r$$

imply the inclusion $c_0 \in A(\Omega)$.

Solutions attracted to Ω simulate a spinodal decomposition when (at the first stage) a transition to a fine-grained structure occurs.

BATES & FIFE [72] reduce Neumann BVP for CHE to an evolutionary equation in a Hilbert space with a densely defined self-adjoint linear part and a smooth nonlinearity. They investigate a dependence between "degree of coarsening" of an equilibrium and degree of its instability. Somewhat more details. Let c_0 be an equilibrium of Neumann BVP for one-dimensional dimensionless CHE

$$c_t = [\omega(c)]_{xx} - c_{xxxx},$$

and λ_n are eigenvalues of the appropriate linearized in c_0 problem numbered in ascending order. The number $l(c_0)$ of intervals of strict monotonicity of the solution c_0 characterizes a degree of coarsening of an alloy (remind that it is a number of transition layers between phases of alloy). Let now c_0 and c_1 are two equilibria, and $\{\lambda_{n0}\}$ and $\{\lambda_{n1}\}$ are corresponding spectra. In BATES & FIFE [72] conditions are specified for the inequality $l(c_0) < l(c_1)$ implies the inequalities $\lambda_{n0} > \lambda_{n1}$ for all n.

BATES & LU [74] prove an analogue of the Grobman-Hartman-Pugh theorem on structural stability for a flow of CHE in a neighborhood of a hyperbolic equilibrium with respect to mass preserving perturbations. They discuss also a decomposition of a flow in a neighborhood of non-hyperbolic equilibrium as product of an infinite dimensional linear flow and a finite-dimensional flow on central manifold.

YONG & ZHENG [151] consider a stabilization problem for solutions of Neumann BVP for CHE. Stabilization of solutions started at any initial function to arbitrary stationary solution through source-discharge control action in the equation is discussed. A problem of minimization of distance between arbitrary stationary solution and any solution in a finite time is also considered.

5.4. Regularization of the equation (3) by the operator $\dfrac{\partial^3 c}{\partial t \partial x^2}$. Stationary solutions and stabilization

The equation

$$c_t = \Delta[\omega(c) + \varepsilon c_t] \tag{104}$$

in a domain $\mathcal{D} \subset \mathbb{R}^n$ can be also considered as a regularization of the equation (3). It was introduced and discussed in NOVICK-COHEN & PEGO [133] in connection with attempt to take into account viscous effects in spinodal decomposition of binary alloys. Simultaneously, this equation ignores an intrephase interactions energy. In the cited paper the following BVP for the equation (104) is considered

$$\frac{\partial}{\partial n}[\omega(c) + \varepsilon c_t]\Big|_{\Gamma} = 0, \tag{105}$$

$$c(x, 0) = c_0, \; x \in \mathcal{D}. \tag{106}$$

By reducing this problem to the nonlinear differential equation

$$c_t = -\varepsilon^{-1}[I - (I - \varepsilon\Delta)^{-1}]\omega(c)$$

in a Banach space E equals to either $L_\infty(\mathcal{D})$, or $C(\overline{\mathcal{D}})$, it is possible to prove existence and uniqueness of solutions in $C^1([0,\infty), E)$ in the case of locally Lipschitz ω (moreover, solutions lie in $C^{k+1}([0,\infty), E)$ if $\omega \in C^{k+1}$, and they are analytic if ω is analytic). Solutions of (104)–(106) keep points of discontinuity of initial function c_0 in the sense that for any $t \in \mathbb{R}$ a function $c(\cdot, t)$ is discontinuous at a point $x \in \mathcal{D}$ iff c_0 is discontinuous at x.

It is easy to show that energy $\mathcal{F}_0 = \int_{\mathcal{D}} W[c(x)]\,dx$ ($W' = \omega$) does not grow along any solution of BVP (104)–(106), and also, solutions save initial average concentration: $\int_{\mathcal{D}} c(x,t)\,dx \equiv \int_{\mathcal{D}} c_0(x)\,dx$.

The BVP (104)–(106) demonstrates a somewhat unusual asymptotic behavior. Obviously, any measurable function $v\colon \mathcal{D} \longrightarrow \mathbb{R}$ such that $\omega[v(x)] \equiv \omega_v =$ const (i.e., step-wise function taking values in the set $\omega^{-1}(\omega_v)$) is a stationary solution of the given equation. Under some additional (and not much restrictive) assumptions, any solution issued from $L_\infty(\mathcal{D})$ is stabilized to such a stationary solution v as $t \to \infty$, in the sense that $\lim_{t\to\infty} c(x,t) = v(x)$ for almost all $x \in \mathcal{D}$, and $\operatorname{mes}\{x \in \mathcal{D}\colon v(x) \in (p_1, p_2)\} = 0$. Moreover, for any stationary solution v satisfying the condition $\omega[v(x)] \geq \alpha > 0$ for almost all $x \in \mathcal{D}$ there exists $\delta > 0$ such that if

$$\|c_0 - v\|_{L_\infty(\mathcal{D})} < \varepsilon \text{ and } \int_{\mathcal{D}} c_0(x)\,dx = \int_{\mathcal{D}} v(x)\,dx, \tag{107}$$

it follows that $\|c(\cdot,t) - v\|_{L_\infty(\mathcal{D})} \longrightarrow 0$ at exponential rate as $t \to \infty$.

Emphasize that in the case when ω_v is not equal to the ordinate of the Maxwell line (see Subsection 3.1.) the stationary solutions described do not provide a minimum to the energy functional $\mathcal{F}_0$. Moreover, a rigidity of restrictions (107) ordered, in particular, coincidence of the discontinuity points c_0 and v, is not a cause of stabilization to a stationary solution that not minimize the energy functional. Roughly speaking, for any stationary solution v a solutions c of (104)–(106) started at initial function c_0 lying in a small L_1-neighborhood of v the following statements hold. First, c is kept in some small L_1-окрестности v. Second, c is close to v in L_∞-norm, for $x \in \mathcal{D}$ such that c_0 is close to v in L_∞-norm. Third, the stationary solution v_c stabilized c is L_1-close to v, and, moreover, v_c and v are close in L_∞-norm for $x \in \mathcal{D}$ such that c_0 and v are L_∞-close and "chemical potentials" ω_v and ω_{v_c} of stationary solutions v and v_c are close.

In Ball, Holmes, James, Pego & Swart [69] the Dirichlet BVP for close to (104) "non-local" one-dimensional ($\mathcal{D} = [0, \pi]$) equation

$$c_t = \left(\int_{\mathcal{D}} c^2(\xi, t)\, d\xi - 1 \right) c_{xx} - \alpha c + \beta c_{xxt}$$

is considered. It is shown that the stationary problem has a countable set of solutions, and each of them has infinite dimensional unstable manifold. Stabilization to a local minimum of energy takes place only in a weak $\overset{\circ}{H}{}^1(\mathcal{D})$ sense.

5.5. Regularization of the equation (2) by operator $\dfrac{\partial^2}{\partial x \partial t}\left[\varphi\left(\frac{\partial}{\partial x}\right)\right]$

Barenblatt, Bertsch, Dal Passo, Prostokishin & Ughi [71,70] suggest the following regularization of the equation (2)

$$\begin{gathered} u_t = [\omega(u_x)]_x + \varepsilon[\varphi(u_x)]_{xt}, \\ (x, t) \in (0, 1) \times (0, +\infty)\, (= \mathcal{D}), \end{gathered} \tag{108}$$

arising as model of mass and heat transfer in picnocline of the ocean (see Subsection 1.4.). It is supposed that $\omega, \varphi \in C^3(\mathbb{R})$, $\omega(0) = 0$, φ is an odd strictly increasing function such that $\gamma = \varphi'(+\infty) \in (0, +\infty)$ and, besides, $|\omega'(p) \leq C\varphi'(p)$ for all $p \in H_0^1(0, 1)$. There are no restrictions on $\omega'(p)$ sign and, therefore, the equation (2) in a described situation can be considered as FBPE (see Fig. 14).

Let us introduce necessary notations. If for $v\colon [0, 1] \longrightarrow \mathbb{R}$ and $x \in (0, 1)$ the limits

$$\lim_{h \to 0} \varphi \left[\frac{v(x + h) - v(x + 0)}{h} \right]$$

and

$$\lim_{h \to 0} \varphi \left[\frac{v(x + h) - v(x - 0)}{h} \right]$$

exist and are equal, then we shall denote their common value by $\Phi[v(x)]$. By Ω we shall denote the function on $[-\gamma, \gamma]$ given by the equality

$$\Omega(p) = \begin{cases} \omega \circ \varphi^{-1} & \text{if} \quad |p| < \gamma, \\ \omega(\pm\infty) & \text{if} \quad p = \pm\infty. \end{cases}$$

By *a solution* of BVP for the equation (108) generated by boundary and initial conditions

$$u_x(0,t) = u_x(1,t) = 0, \ t \in (0, +\infty), \qquad (109)$$

$$u(x,0) = u_0(x), \ x \in (0,1), \qquad (110)$$

we shall mean a function $u: \mathcal{D} \longrightarrow \mathbb{R}$ such that its restriction onto $\mathcal{D} = (0,1) \times (0,T]$ (which we shall denote by the same letter u) satisfies the following conditions:

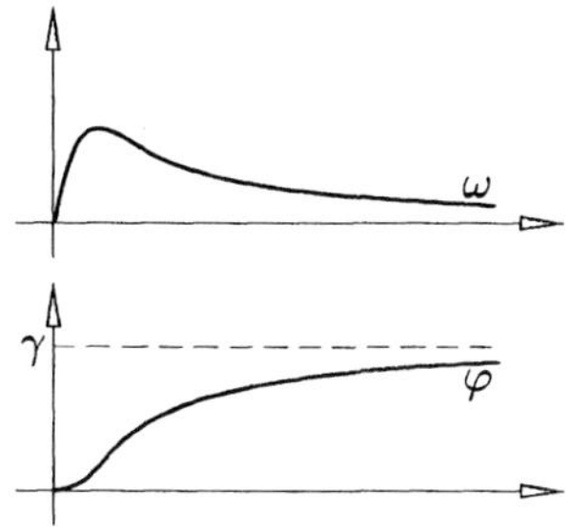

Fig. 14

i) $u \in L_\infty(\mathcal{D}) \cap L_\infty([0,T], BV[0,1])$, $u_t \in L_2(\mathcal{D})$;□

ii) the function $w(x,t) = \Phi[u(x,t)]$ is defined in $\mathcal{D}$, and, moreover, $w \in C(\mathrm{cl}\,\mathcal{D}) \cap L_\infty([0,T], \overset{\circ}{H}{}^1[0,1])$ and $w_t \in L_2([0,T], H^1[0,1])$;□

iii) the functions u, w satisfy the equation

$$u_t = [\Omega(w)]_x + \varepsilon w_{xt}$$

in the sense of $L_2(\mathcal{D})$, and the initial condition

$$u(x,0) + \varepsilon w_x(x,0) = u_0(x) + \varepsilon[\Phi(u_0)]'.$$

In the cited works a solvability of the BVP (108)–(110) in the described class of solutions is proved. Besides, it is shown that the additional condition

$$\left| \left(\frac{\omega'(p)}{\varphi'(p)} \right) \cdot \frac{1}{\varphi'(p)} \right| \le C \text{ for all } p \in \mathbb{R}$$

guarantees uniqueness of solutions.

Solutions of (108)–(110) demonstrate somewhat unusual asymptotic behavior, evolving to step functions; this, in the authors' opinion, simulates occurrence of layered structures in the ocean picnocline. Let the above restriction holds and, moreover, $\omega(+\infty) = 0$, $\omega(p) < 0$ for $p < 0$ and u_0 does not decrease on $[0,1]$.

Theorem. *If $w(x_0, t_0) = \gamma$ for some $x_0 \in (0,1)$, it follows that $w(x_0,t) = \gamma$ for all $t > t_0$, and $u(x_0{+}0, t)$ does not decrease and $u(x_0{-}0,t)$ does not increase on $(t_0, +\infty)$.*

Theorem. *There exists $u_0 \in C^1[0,1]$ such that the appropriate solution of the problem (108)–(110) is not continuous in $\mathcal{D}$.*

Theorem. *For any solution u of the problem (108)–(110) there exists a function $v \in BV[0,1]$ such that $v'(x) = 0$ almost everywhere on $[0,1]$ and $u(x,t) \longrightarrow v(x)$ as $t \to +\infty$ for almost all $x \in [0,1]$.*

As indicated above, the equation (108) is result of a linearization with respect to a small delay in the diffusion equation with diffusion coefficient dependent on gradient of solution in the past

$$u_t = [\nu(u_x^\tau)u_x]_x, \tag{111}$$

where $u_x^\tau(x,t) = u(x,t-\tau)(\tau > 0)$. It should be emphasized that for (111) in contrast to the equation (108) an effect of formation of discontinuities of solution of BVPs for the finite time is not observed AKHMEROV & SAPRYKIN [58].

6. Passage to the limit in regularizators

In this Section various passages to the limit with respect to a small parameter in the regularizing trems are described. In particular, the results provide partially answers to such a problem as notions solution of ill-posed BVP for FBPEs.

6.1. Limiting measure-valued solutions

We assume here, that $p_1 = -1$, $p_2 = 1$, $\omega(p_1) = 1$, $\omega(p_2) = -1$, and that the functions f_1', f_2', f_3' (see the Subsection 2.1.) are linearly independent on any open subset of the interval $(-1,1)$. We shall consider the following BVP

$$u_t = \Delta\omega(u) + \varepsilon\Delta u_t,\ (x,t) \in \mathcal{D}, \tag{112}$$

$$\varepsilon u_t + \omega(u) = 0,\ (x,t) \in \partial\mathcal{D} \times [0,T], \tag{113}$$

$$u(x,0) = u_0(x),\ x \in \mathcal{D}. \tag{114}$$

As in NOVICK-COHEN & PEGO [133] (see Subsection 5.4.), it is easy to show uniqueness and existence of solutions of the problem (1'2), (113), (114) in the class $u \in L_\infty(\mathcal{D})$, $u_t \in L_\infty([0,T], W_p^2(\mathcal{D}))$ $(1 \le p < \infty)$ for any function $u_0 \in L_\infty(\mathcal{D})$.

It turns to be out (see PLOTNIKOV [36,37]) that strong L_p-limits as $\varepsilon \longrightarrow 0$ of solutions of the problem (112), (113), (114) generate a subclass of atomic measure-valued solutions of the BVP

$$u_t = \Delta\omega(u),\ (x,t) \in \mathcal{D}, \tag{115}$$

$$\omega(u) = 0,\ (x,t) \in \partial\mathcal{D} \times [0,T], \tag{116}$$

$$u(x,0) = u_0(x),\ x \in \mathcal{D} \tag{117}$$

(the equation (115) is a multidibensional version of (3)). To describe this subclass we consider the following problem (as in PLOTNIKOV [37], we shall refer it as *problem A*): find such functions $u, \lambda_i \in L_\infty(\mathcal{D})$ $(i = 1,2,3)$ and $v \in L_\infty(\mathcal{D}) \cap L_2([0,T], \overset{o}{H}{}^1(\mathcal{D}))$ that

$$\sum_{i=1}^{3} \lambda_i(x,t) = 1,\ \lambda_i(x,t) \ge 0,\ (x,t) \in \mathcal{D},$$

$$\lambda_1(x,t) = 1, \text{ for } v(x,t) < -1, \text{ and } \lambda_3(x,t) = 1, \text{ for } v(x,t) > 1,$$

$$u(x,t) = \sum_{i=1}^{3} \lambda_i(x,t) f_i[v(x,t)], \ (x,t) \in \mathcal{D}, \tag{118}$$

$$u_t = \Delta v, \ (x,t) \in \mathcal{D}, \quad u(x,0) = u_0(x), \ x \in \mathcal{D},$$

$$v(x,t) = 0, \ (x,t) \in \partial\mathcal{D} \times [0,T],$$

and, finally, for any non-decreasing function $h \in C^1(\mathbb{R})$

$$\mathcal{H}_t(x,t) - h[v(x,t)]u_t(x,t) \le 0, \ (x,t) \in \mathcal{D}, \tag{119}$$

where

$$\mathcal{H} = \sum_{i=1}^{3} \lambda_i H[h_i(v)], \ H(p) = \int_0^p h[\omega(s)]\, ds.$$

The equation (118) and the inequalities (119) are understood in an integral sense:

$$\int_{\mathcal{D}} (u\xi_t - \nabla v \nabla \xi)\, dx\, dt = \int_{\mathcal{D}} u_0 \xi(x,0)\, dx = 0$$

for all smooth functions ξ equal to zero on $\partial\mathcal{D} \times [0,T]$, and (hereafter, g denotes a primitive of a function h)

$$\int_{\mathcal{D}} [\mathcal{H}\xi_t + g(v)\Delta\xi - h'(v)|\nabla v|^2 \xi]\, dx\, dt \ge 0$$

for all smooth non-negative ξ.

Let now u_ε be a solution of the problem (112), (113), (114), and $v_\varepsilon = \omega(u_\varepsilon)$. It is easy to prove (see above) that there exist a sequence ε_n and functions $u, v \in L_\infty(\mathcal{D})$ such that

$$u_{\varepsilon_n} \longrightarrow u, \text{ and } v_{\varepsilon_n} \longrightarrow v \ \ast\text{-weakly in } L_\infty(\mathcal{D}).$$

It appears, that $v_{\varepsilon_n} \longrightarrow v$ strongly in $L_\infty(\mathcal{D})$, and, besides, there exist $\lambda_i \in L_\infty(\mathcal{D})$ $(i = 1,2,3)$ such that the set (u, v, λ_i) is the solution of the problem A!

It is easy to see that the family of atomic probabilistic measures

$$V(x,t) = \sum_{i=1}^{3} \lambda_i(x,t) \delta[\tau - f_i(v(x,t))]$$

determines the measure-valued solution $(x,t) \mapsto V(x,t)$ of the problem (115), (116), (117). Besides, the pair (u,v), obviously, can be treated as the solution of the differential inclusion

$$u_t = \Delta v, \ v \in \Omega(u)$$

(cf. with Subsection 3.5.).

Moreover, in some sense, solutions of the problem A one can interpet as hysteresis one, namely, if $\omega'(p_1), \omega'(p_2) \neq 0$ and $\operatorname{vrai\,sup} v(x,t) < p_2$ on $\mathcal{D} \times [t_0, t_1]$, it follows that for almost all $x \in \mathcal{D}$ the function $\lambda_1(x,\cdot)$ is not decreasing on $[t_0, t_1]$. Similarly, the function $\lambda_3(x,\cdot)$ is not decreasing on $[t_0, t_1]$, if $\operatorname{vrai\,sup} v(x,t) > p_1$ on $\mathcal{D} \times [t_0, t_1]$. If $u_0(x) < p_0$ (respectively, $u_0(x) > p_3$), then $\lambda_1(x,t) \equiv 1$ (respectively, $\lambda_3(x,t) \equiv 1$) almost everywhere on $\mathcal{D}$.

6.2. CHE and Hele-Show problem

At the first stage of evolution solutions of the CHE quickly pass to the two-components state of fine-grained solution, i.e., in the domain $\mathcal{D}$ of $u(\cdot,t)$ two subareas $\mathcal{D}_+$ and $\mathcal{D}_-$ can be allocated, in which $u(x,t) \approx 1$ and $u(x,t) \approx -1$ (it is assumed that $p_0 = -1$ and $p_1 = 1$), respectively, and $\operatorname{mes}[\mathcal{D} \setminus (\mathcal{D}_+ \cup \mathcal{D}_-)]$ is small. At the following stage, the areas $\mathcal{D}_+$ and $\mathcal{D}_-$ slowly evolve trying to minimize the surface interaction energy. To describe the second stage, it is convenient to replace ε by ε^2 and to scale time by change t by εt. Then CHE will be rewritten in the form

$$u_t = \Delta[\varepsilon^{-1}\omega(u) - \varepsilon \Delta u]. \tag{120}$$

It turns out that solutions of Neumann BVP for this equation tends to solutions of the Hele-Show free boundary problem as $\varepsilon \longrightarrow 0$. The latter problem is a problem on finding of a function v in $\mathcal{D}$ and a free boundary $\Gamma \subset \mathcal{D}$, $\Gamma = \bigcup_{0 \le t \le T} (\Gamma_t \times \{t\})$ satisfying the following equations:

$$\begin{gathered}
\Delta v = 0, \; t \in [0,T], \; x \in \mathcal{D} \setminus \Gamma_t, \\
\frac{\partial v}{\partial n} = 0, \; (x,t) \in \partial \mathcal{D} \times [0,T], \\
v = \sigma \kappa_{\Gamma_t}, \; t \in [0,T], \; x \in \Gamma_t, \\
V = \frac{1}{2} \left[\frac{\partial v}{\partial n} \right] \Big|_{\Gamma_t}, \; t \in [0,T], \; x \in \Gamma_t, \\
\Gamma_0 \text{ is given;}
\end{gathered} \tag{121}$$

where κ_{Γ_t} is a main curvature of Γ_t, $\sigma = \text{const}$, $V = V(x,t)$ is a normal velocity of surface Γ at (x,t), $\left[\frac{\partial v}{\partial n} \right] \Big|_{\Gamma_t}$ is a jump of a function $\frac{\partial v}{\partial n}$ on Γ_t. It turns out that for any solution of the Hele-Show problem there exists a solution of a special BVP for the equation (120) approximating it as $\varepsilon \longrightarrow 0$. Namely, the following theorem holds.

Theorem (ALIKAKOS, BATES & CHEN [64]). *Let $\mathcal{D}$ be a domain with smooth boundary and Γ_0 be a smooth hypersurface in $\mathcal{D}$. Let the problem (121) has a smooth solution (v, Γ) such that $cl\Gamma_t \subset \mathcal{D}$ for all $t \in [0,T]$. Then for all sufficiently small $\varepsilon > 0$ there exist uniformly bounded (with respect to ε)*

functions $u_0^\varepsilon \in C^\infty(\mathcal{D})$ *and* $J^\varepsilon \in C^\infty(cl(\mathcal{D}))$ *such that started at* u_0^ε *solution* u^ε *of the BVP for the equation* (120) *with boundary conditions*

$$\frac{\partial}{\partial N}[\varepsilon^{-1}\omega(u)_\varepsilon \Delta u] = 0, \ (x,t) \in \partial\mathcal{D},$$

$$\frac{\partial u}{\partial n}\varepsilon^4 J^\varepsilon(x,t, \ (x,t) \in \partial\mathcal{D},$$

has the following properties:

$$\lim_{\varepsilon\to 0}[\varepsilon^{-1}\omega(u)_\varepsilon \Delta u] = -v, \ \ (x,t) \in \partial\mathcal{D},$$

and

$$\lim_{\varepsilon\to 0} u^\varepsilon(x,t) = \{+1 \text{ if } (x,t) \in \mathcal{D}^+, -1 \text{ if } (x,t) \in \mathcal{D}^-,$$

where $\mathcal{D}^+$ *and* $\mathcal{D}^-$ *are the interior and exterior of the domain with boundary* Γ. *If the function* ω *is linear, it follows that* $J^\varepsilon(x,t) \equiv 0$ *in neighborhoods of the points* ± 1.

A formal matched asymptotic expansion in terms of degrees of ε for solutions of specified BVP was constructed earlier in Pego [136]. This expansion leads to the similar statement. The Dirichlet BVP is also considered.

6.3. Stationary solutions of the Cahn-Hilliard in variational statement

We consider a problem of description of a steady state of two-phase solutions in view of energy of interphase interaction (see Subsection 1.2.). In variational posing this is a problem of minimization of functional

$$\mathcal{F}_\varepsilon(c) = \int_0^1 \left\{W[c(x)] + \frac{\varepsilon}{2}[c'(x)^2]\right\} dx \longrightarrow \min \tag{122}$$

under the restrictions (38)–(39). Passing to Euler-Lagrange equations leads to the BVP for equation

$$\varepsilon c'' = \omega(c) - \lambda \tag{123}$$

with boundary conditions (38)–(39).

A function (40) will be called *a periodic solution* of the BVP (37)–(39), if $\mathcal{D}_A$ and $\mathcal{D}_B$ consist of intermittent intervals of equal length. It is easy to see that the problem (37)–(39) has a countable number of periodic solutions.

A solution c will be called *a regular solution* of this problem, if there exist sequences $\varepsilon_n \longrightarrow 0$, $\lambda_n \longrightarrow \lambda_0$, $m_n \longrightarrow m$ and a sequence c_{ε_n} of solutions of the problem (123), (38), (39) such that

$$||c_{\varepsilon_n} - c||_{L_1[0,1]} \longrightarrow 0 \text{ as } n \to \infty.$$

Theorem. *The set of periodic solutions of the problem* (37)–(39) *coincides with the set regular solutions.*

The complete proof of this intuitively clear theorem is much cumbersome and forms a greater part of the article CARR, GURTIN & SLEMROD [84]. It amounts to the accurate analysis of a phase portrait of the equation (123). Certainly, the point is that this equation is singularly perturbed.

Let us denote a periodic solution of the problem (37)–(39) having of N jumps by c^N (in particular, c^1 is a simple solution), and an appropriate smooth solution of the problem (123), (38), (39) approximating a solution c^N as $n \to \infty$ by $c^N_{\varepsilon_n m_n}$.

Theorem. *The function $c^N_{\varepsilon_n m_n}$ for $N = 1$ and sufficiently large n is a global minimum of energy functional $\mathcal{F}_{\varepsilon_n}$. For $N > 1$ these functions are not even local minima of $\mathcal{F}_{\varepsilon_n}$.*

The statement of the theorem follows from the following asymptotic representation of energy

$$\mathcal{F}_{\varepsilon_n}(c^1_{\varepsilon_n m_n}) = 2[\lambda_0 m_n + W(c^1_{\varepsilon_n m_n}) - \lambda_0 c_A] + \varepsilon S + O(e^{-C/\varepsilon}),$$

where

$$s = 2\int_{c_A}^{c_B} [W(s) - W(c_B) - \lambda_0(s - c_B)]^{\frac{1}{2}}\, ds.$$

6.4. On a minimum surface of interaction of components in binary alloys

The statement described is a multidimensional extension of previous subsection. Let Ω be an open bounded set in $\mathbb{R}^m$ with Lipschitz boundary. Let us consider the following stationary BVP for (95)

$$\varepsilon\Delta^2 c = \Delta\omega(c), \;\; x \in \Omega,$$

$$\frac{\partial c}{\partial n} = \frac{\partial}{\partial n}[\omega(c) - \varepsilon\Delta c] = 0, \;\; x \in \partial\Omega \tag{124}$$

where Δ is the Laplas operator, and n is an external normal to Ω). This problem is equivalent to the problem of minimization of energy functional

$$\mathcal{F}_\varepsilon(c) = \int_\Omega \left\{W[c(x)] + \frac{\varepsilon}{2}|\nabla c(x)|^2\right\} dx \longrightarrow \min \tag{125}$$

under the restriction

$$\int_\Omega c(x)\, dx = m \tag{126}$$

and the boundary conditions (124). The degenerate BVP ($\varepsilon = 0$)

$$\Delta\omega(c) = 0, \;\; x \in \Omega,$$

$$\frac{\partial c}{\partial \nu} = 0, \ x \in \partial\Omega$$

corresponds to the minimization problem of degenerate energy functional

$$\mathcal{F}_\varepsilon(c) = \int_\Omega W[c(x)]\,dx \longrightarrow \min \tag{127}$$

with the same restriction (126). Minimization is progressed on the space $L_1(\Omega)$ (in doing so the functional $\mathcal{F}_\varepsilon(c)$ on functions from $L_1(\Omega)\backslash H_1(\Omega)$ is assumed to be equal to $+\infty$).

A mention has already been made (see Subsection 3.1.) that solutions of the problem (127) are step-wise functions of the form (40), where $\mathcal{D}_A$ and $\mathcal{D}_B$ are measurable nonintersecting subset of Ω such that $\mathcal{D}_A \cup \mathcal{D}_B = \Omega$ and

$$\text{mes}\mathcal{D}_A = \frac{c_B \text{mes}\Omega - m}{c_B - c_A}.$$

The number

$$P_\Omega(E) = \sup\left\{\int_\Omega \chi_E(x)\sum_{i=1}^m (\varphi_i)_{x_i}(x)\,dx \colon \varphi \in \overset{\circ}{C}{}^\infty(\Omega, \mathbb{R}^m),\ \|\varphi\| \le 1\right\},$$

where χ_E is characteristic function of a set E, is called to be *a perimeter in Ω of measurable set $E \subset \mathbb{R}^m$.*

It turns out that a function is a limit of solutions of the problem (125)–(126) as $\varepsilon \longrightarrow 0$ only if it is a solutions of the problem (127)–(126) which components $\mathcal{D}_A$ and $\mathcal{D}_B$ of minimal in Ω perimeter. More exactly, the following theorem holds.

Theorem (Modica [125]). *Let $m \in [c_A\text{mes}\Omega, c_B\text{mes}\Omega]$, $\varepsilon_n \longrightarrow 0$ as $n \to \infty$, $\varepsilon_n > 0$ and c_n be a solution of the problem (126) for $\varepsilon = \varepsilon_n$. Let c_n converges in $L_1(\Omega)$ to some function c_0 as $n \to \infty$.*

Then $W[c_0(x)] \in \{c_A, c_B\}$ for almost all $x \in \Omega$; the set $\mathcal{D}_A$ ($= \{x \in \Omega\colon c_0(x) = c_A\}$) has a minimal perimeter, i.e., it is a solution of the optimization problem

$$P_\Omega(E) \longrightarrow \min$$

under the restrictions

$$E \subset \Omega,\ E \text{ is measurable},\ \text{mes}E = \frac{c_B\text{mes}\Omega - m}{c_B - c_A};$$

and, at last, $\lim_{n\to\infty} \frac{1}{\sqrt{\varepsilon_n}}\mathcal{F}_{\varepsilon_n}(c_n) = 2P_\Omega(\mathcal{D}_A)\int_{c_A}^{c_B}\sqrt{W(s)}\,ds.$

6.5. Limiting dynamics of "bubbles" and "drops" in CHE

In connection with complication of a structure of attractor $\mathring{A}_\varepsilon$ of CHE as $\varepsilon \to 0$ (unlimited growth of number of steady states, "degregation" of their stability properties, unlimited growth of the attractor's dimension) a number of authors (see the references below) offer to investigate behaviour (as $\varepsilon \to 0$) of the sets $\mathcal{C}_E \stackrel{\text{def}}{=} \{c: \mathcal{F}_\varepsilon(c) \leq E\}$ by suitable choice of an energy level E. ALIKAKOS & FUSCO [63] assume that in general case for all $\varepsilon > 0$ sufficiently small one can find such $E_0, \ldots, E_{\dim(\mathring{A}_\varepsilon)}$ that

$$\dim \mathcal{C}_E = i \text{ for } E_i \leq E < E_{i+1}.$$

The authors believe that structure of sets $\mathcal{C}_{E_i}$ as $\varepsilon \to 0$ is a lot of easier, than structure of $\mathring{A}_\varepsilon$.

In ALIKAKOS, BRONSARD & FUSCO [61] and ALIKAKOS, CHEN & FUSCO [62] the following two theorems are proved describing behaviour of an elementary interfaces ("moving bubbles" and "spreading along a border drops") as $\varepsilon \to 0$ (they are cited here according to ALIKAKOS & FUSCO [63]).

It is assumed that $\omega(p) = p(p^2 - 1)$. For any $a > 0$ we shall denote the set $\{x \in \mathcal{D}: \rho(x, \partial\mathcal{D}) > a\}$ by $\mathcal{D}_a$. Let b be such that $\mathcal{D}_b \neq \emptyset$ and $\delta > 0$ be so small that $\mathcal{D}_{b+\delta} \neq \emptyset$. For any $\zeta \in \operatorname{cl} \mathcal{D}_{b+\delta/2}$ define the function $s_\zeta: \mathcal{D} \to \mathbb{R}$ ("limiting bubble") by the equality

$$s_\zeta(x) = \begin{cases} -1 & \text{for} \quad \|x - \zeta\| < b, \\ 1 & \text{for} \quad \|x - \zeta\| \geq b. \end{cases}$$

Finally, let $\rho_\zeta = \rho(\zeta, \partial\mathcal{D}) - b$.

Theorem (ALIKAKOS, BRONSARD & FUSCO [61]). *There exist $\varepsilon_0 > 0$ and $C_1, C_2 > 0$ such that for any $\zeta_0 \in \mathcal{D}_{b+\delta}$ and $\varepsilon \in (0, \varepsilon_0]$ there is a solution c_ε of BVP (95)–(96) such that*

i) $\|c_\varepsilon(\cdot, 0) - s_{\zeta_0}\|_{L_1(\mathcal{D})} \leq C_1\varepsilon$;

ii) either

$$\sup_{t \geq 0} \inf_{\zeta \in \mathcal{D}_{b+\delta}} \|c_\varepsilon(\cdot, t) - s_\zeta\|_{L_1(\mathcal{D})} \leq C_1\varepsilon,$$

or there exists $T > 0$ such that

$$\inf_{\zeta \in \partial\mathcal{D}_{b+\delta}} \|c_\varepsilon(\cdot, T) - s_\zeta\|_{L_1(\mathcal{D})} \leq C_1\varepsilon,$$

and, moreover, $T > \exp\left(\dfrac{C_2\rho_{\zeta_0}}{\varepsilon}\right)$.

For any $\zeta_0 \in \partial\mathcal{D}$ we denote by $\tilde{\zeta}: [0, \infty) \to \partial\mathcal{D}$ a solution of Cauchy problem

$$\frac{d\zeta}{d\tau} = \beta \operatorname{grad} \kappa_{\partial\mathcal{D}}(\zeta), \ \tau \geq 0,$$

$$\zeta(0) = \zeta_0,$$

where $\kappa_{\partial\mathcal{D}}(\zeta)$ is the mean curvature of $\partial\mathcal{D}$ at the point ζ, and β is a positive constant.

Theorem (Alikakos, Chen & Fusco [61]). *There exists $\varepsilon_0 > 0$ such that for any $\zeta_0 \in \partial\mathcal{D}$ and any $\varepsilon \in (0, \varepsilon_0]$ there is a solution c_ε of BVP (95)–(96) such that for all $\tau > 0$*

$$\lim_{\varepsilon\to 0} c_\varepsilon(x, \varepsilon^{-1}\tau) = \begin{cases} 1 & \textit{for} \quad x \in \operatorname{cl}\mathcal{D}\backslash\{\tilde{\zeta}(\tau)\}, \\ -1 & \textit{for} \quad x = \tilde{\zeta}(\tau). \end{cases}$$

6.6. Regular periodic traveling waves of the equation (1)

We consider the equation (1) regularized by the small dispersion:

$$u_t + uu_x = [\omega(u_x)]_x + \varepsilon u_{xxx}, \tag{128}$$

where ε is a small positive parameter. In this and the following subsections we are interested in traveling waves of the equation (1) which are limits of traveling waves of the regularized equation as $\varepsilon \to 0$.

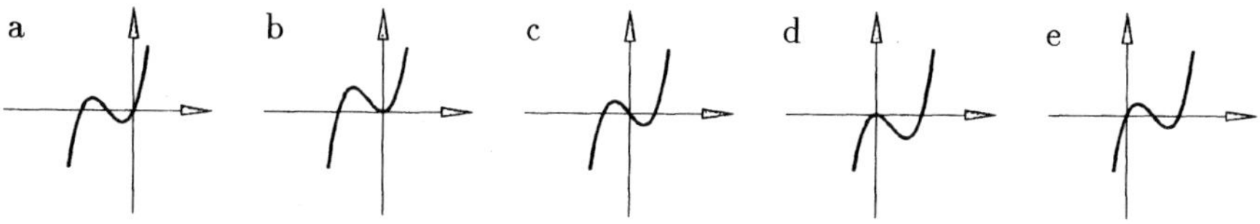

Fig. 15

We shall distinguish five various cases represented on Fig. 15, depending on arrangement of local extrema of the function ω relatively to zero.

As well as in Subsection 3.6., the problem on traveling waves of the equation (128) is equivalent to the problem on solutions of the ordinary differential equation

$$\varepsilon v'' = g_c(v) - \omega(v'),$$

which, in turn, is equivalent to the singularly perturbed system of the first order equations

$$v' = w, \tag{129}$$

$$w' = \varepsilon^{-1}[g_c(v) - \omega(w)]. \tag{130}$$

A function $u(x,t) = v(x - \alpha t) + \alpha$ generated by v-component of a smooth periodic solutions of the system (129)–(130) is called *a periodic traveling wave of the equation* (128). A periodic traveling wave of the equation (1) which is also a limit in PW_p^1 (see Subsection 3.7.) of periodic traveling waves of the equation (128) as $\varepsilon \to 0$ will be called *regular*.

Theorem (Akhmerov [55,56]). *In the cases (a), (b), (d) and (e) represented on Fig. 15 the equation (1) has no regular periodic traveling waves. In the case (c) it has a unique regular periodic traveling wave for $c < \omega(p_2)$ and has no such waves for $c \geq \omega(p_2)$.*

The proof of this theorem is standard for the theory of singularly perturbed systems of ordinary differential equations (cf. with relaxation oscillations of

Van der Pol oscillator). Fig. 16 represents a trajectory of the system (129)–(130) corresponding to a regular periodic wave of the equation (128) for small ε.

An existence some of solutions described is noted in YANENKO & SOLOV'EV [52].

We shall note here that the solvability of a BVP for (128) was proven in PODGAEV [39].

6.7. Forward backward parabolic periodic and shock transition traveling waves of the equation (1)

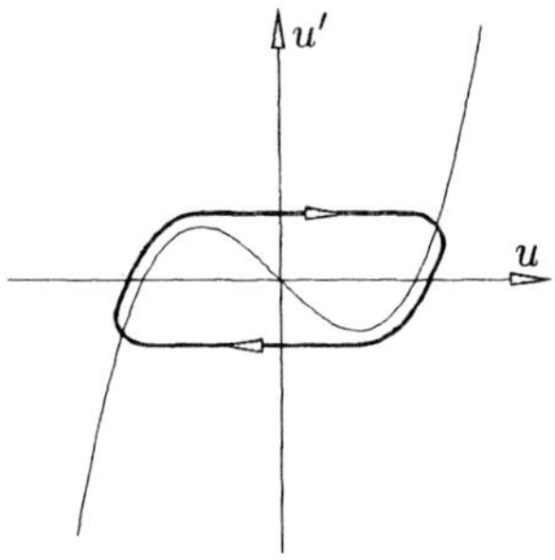

Fig. 16

We note that discussed in Subsections 3.1., 3.7., 4.2., 4.4., 4.5. 4.10. solutions "spent almost all time in region of forward parabolicity" in the sense that

$$\text{mes}\{(x,t)\colon \omega'[u_x(x,t)] \le 0\} = 0.$$

At the same time, in some real systems processes with negative diffusion coefficient are observed. Therefore solutions of the equation (1) for which

$$\begin{aligned} &\text{mes}\{(x,t)\colon \omega[u_x(x,t)] > 0\} > 0, \\ &\text{mes}\{(x,t)\colon \omega'[u_x(x,t)] < 0\} > 0 \end{aligned} \tag{131}$$

are of interest. We shall refer to such solutions as *forward backward parabolic solutions.* These solutions can be easily constructed, for example, in the case of the stationary Dirichlet BVP (see AKHMEROV [5,6]). From our point of view, more intriguing is a question: is it possible to get such a solution as a result "discarding of small regularizing terms"? It turns out, that in degenerate cases when $\omega'(0) = 0$ (see (b) and (d) on Fig. 15), solutions satisfying (131) can be obtained as a result of matched passing to the limit with respect to two small parameters, namely, small dispersion and small "shift" in a nonlinear diffusion coefficient ω'. True, this way also not permits to allocate a unique solution. But the set of such solutions is a continuous curve, instead of "catastrophically rich" set of all generalized solutions (see Subsections 3.4., 4.3.).

In parallel with the equation (1) we consider the equation

$$u_t + uu_x = [\omega_\xi(u_x)]_x + \varepsilon u_{xxx} \tag{132}$$

with two small parameters ξ and ε; where $\omega_\xi\colon \mathbb{R} \longrightarrow \mathbb{R}$ is a function defined by the formula

$$\omega_\xi(p) = \omega(p+\xi).$$

We have taken such a form of ω_ξ for simplicity only. Actually, it is possible to consider essentially more general dependence ω on ξ.

A periodic traveling wave u of the equation (1) such that there exist a function $\xi: \mapsto \xi(\varepsilon)$ and periodic traveling waves $u_{\varepsilon\xi(\varepsilon)}$ of the equation (132) such that $u_{\varepsilon\xi(\varepsilon)} \longrightarrow u$ in PW^1_p as $\varepsilon \to 0$ will be called *limiting*.

We consider only the case $p_1 = 0$. The case $p_2 = 0$ is similar.

Singularly perturbed system of ordinary differential equations

$$v' = w, \tag{133}$$

$$w' = \varepsilon^{-1}[g_c(v) - \omega_\xi(w)]. \tag{134}$$

is equivalent to the equation (132). The direction field of this system corresponding to the case $c < \omega(p_2)$ is depicted on Fig. 17. Solution $u_{\varepsilon\xi(\varepsilon)}$,

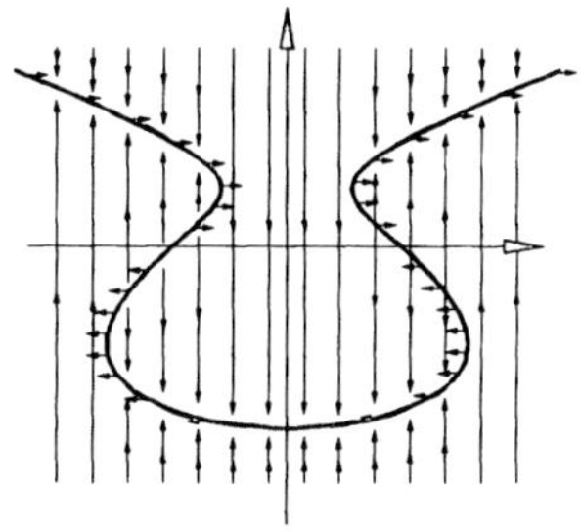

Fig. 17

which will meet to limiting periodic traveling forward backward parabolic wave, completely correspond to so called solutions-duck of singularly perturbed differential equations opened by Strasbourg group of professor J.Reeb with a non-standard analysis tools.

A limiting periodic traveling wave generated by the cycle of (133)–(134) presented on Fig. 18a), as well as, a solution itself, we shall call traditionally *a duck with a head*, and the cycle presented on Fig. 18b) *a duck without a head*. A number p will be referred as *a neck of the duck*.

Theorem (AKHMEROV [55,56]). *For any $c < \omega(p_2)$ and $p \in (0, p_2)$ there exists a unique limiting periodic traveling wave of the equation (1) generated by a duck with the neck p, either with head, or without head.*

The theorem can be proved by the non-standard analysis tools (see, f.e., DAVIS [14], ROBINSON [140]) immersing the system (133)–(134) into an appropriate non-standard one and discussing behavior of phase trajectories in halo of the cycle Γ in the spirit of BENOIT, CALLOT, DIENER & DIENER [91].

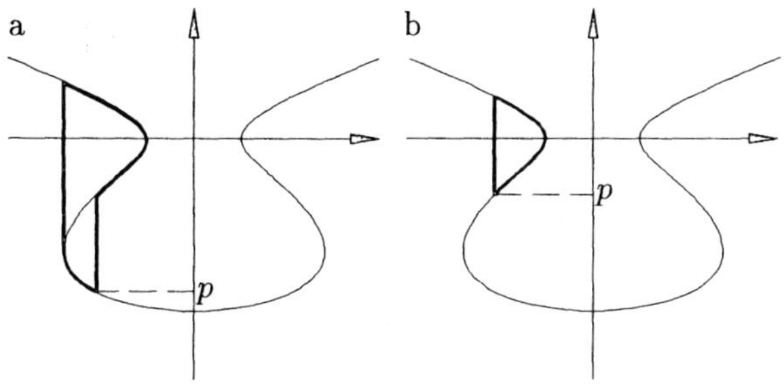

Fig. 18

Closely analogous statements take place for traveling waves of shock transition type AKHMEROV [56]. We shall restrict our citation only by formulations of some simplest theorems and figures. Regular and limiting traveling on a zero and oscillating background shock transition wave (see Subsection 3.6.) are defined just in the same way as in periodic case.

We assume that in the cases (b) and (d) on Fig. 15 $\omega''(0) \neq 0$ and $\omega'''(0) \neq 0$.

Theorem. *In the case (c) on Fig. 15, as well as in the cases (b) if* $\omega'''(0) < 0$ *and (d) if* $\omega'''(0) > 0$, *the equation* (1) *has no regular shock transition waves traveling on zero background. In the rest cases it has a unique such wave.*

Theorem. *In the cases (a), (b), (d), and (e) the equation* (1) *has no regular shock transition waves traveling on oscillating background. In the case (c) it has a unique such wave.*

The limiting forward backward parabolic traveling shock transition waves exist only in the cases (b) and (d) presented on Fig. 15. More precisely, the following theorem holds.

Theorem. *If* $\omega'(0) \neq 0$, *the equation* (1) *has no limiting forward backward parabolic traveling on a zero or oscillating background shock transition waves.*

Thus, it is supposed below that $\omega'(0) = 0$.

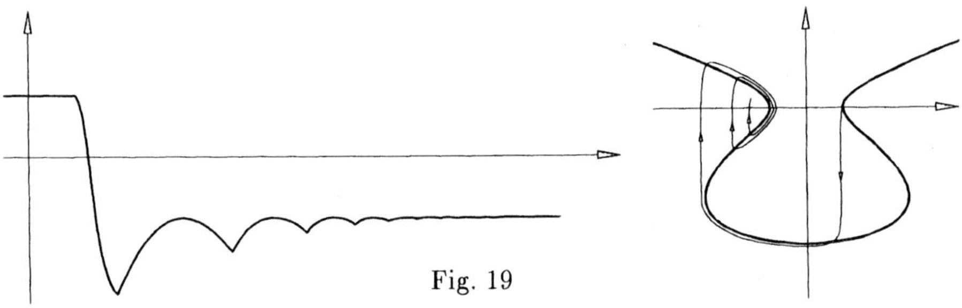

Fig. 19

Theorem. *In the cases (b) if* $\omega'''(0) < 0$ *and (d) if* $\omega'''(0) > 0$ *the equation* (1) *has a continuum of limiting forward backward parabolic traveling on zero background shock transition waves (case (b) see on Fig. 19). In the other cases it has no such solutions.*

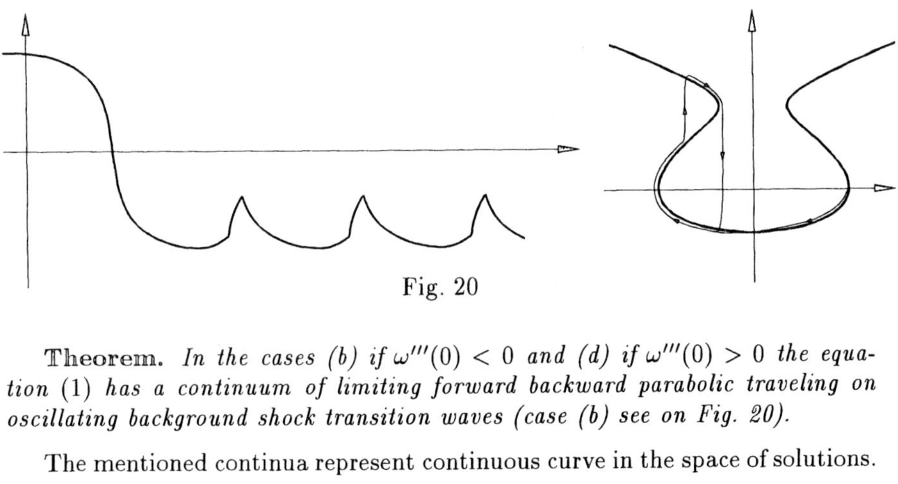

Fig. 20

Theorem. *In the cases (b) if* $\omega'''(0) < 0$ *and (d) if* $\omega'''(0) > 0$ *the equation* (1) *has a continuum of limiting forward backward parabolic traveling on oscillating background shock transition waves (case (b) see on Fig. 20).*

The mentioned continua represent continuous curve in the space of solutions.

References

1. R.R. Akhmerov, On structure of a solution set of a boundary value problem for one-dimensional stationary equation of variable type, Numer. Methods in Mech. Continuum Media (Численные методы механики сплошной среды), Novosibirsk, 15, No. 5, 1984, pp. 20–30 (in Russian).

2. R.R. Akhmerov, Neumann problem for a differential equation with coefficient at the leading derivative of variable sign, Numer. Methods in Mech. Continuum Media (Численные методы механики сплошной среды), Novosibirsk, 16, No. 1, 1985б pp. 3–11 (In Russian).

3. R.R. Akhmerov, On stationary one-dimensional equation with a sign-changing coefficient of viscosity, In The Fourth Int. Conf. on Boundary and Interior Layers, Novosibirsk, July, 1986, pp. 7–11.

4. R.R. Akhmerov, On structure of a set of solutions of Dirichlet boundary value problem for stationary one-dimensional forward-backward parabolic equation, Nonlinear Anal.: Theory, Methods and Appl., 11, No. 11, 1987, pp. 1303–1316.

5. R.R. Akhmerov, On periodic travelling waves of equations with viscosity coefficient of variable sign, Nonlinear Anal.: Theory, Methods and Appl., 13, No. 7, 1989, pp. 803–817.

6. R.R. Akhmerov, Travelling waves of the Burger and Korteweg-de Vries-Burger equation with viscosity coefficient of variable sign, Russian J. of Theor. and Appl. Mech., 1, No. 1, 1991, pp. 27–37.

7. R.R. Akhmerov, On a set of solutions of a nonlinear forward backward parabolic equation, Russian J. of Numer. Anal. and Math. Modelling, 8, No. 6, 1993, pp. 453–459.

8. R.R. Akhmerov, M.I. Kamenskii, A.S. Potapov, A.E. Rodkina, B.N. Sadovskii, Measures of Noncompactness and Condensing Operators, volume 55 of Operator Theory, Birkhäuser Verlag, Basel, e.a., 1992.

9. R.R. Akhmerov, V.A. Novikov, On a problem of solvability of a second order differential equation with coefficient at the leading derivative of variable sign, Preprint, Inst. Pure Appl. Mech., Siberian Division of USSR Acad. Sci., No. 39, 1981 (in Russian).

10. R.R. Akhmerov, V.A. Novikov, Investigation of solutions of a differential equation not solved for the leading derivative, Soviet Math. Dokl., 25, No. 3, 1982, pp. 551–553.

11. R.R. Akhmerov, A.I. Saprykin, Forward backward parabolic equations and delay partial differential equiations, (in preparation), 1995.

12. Vasilios Alexiades, Elias C. Aifantis, On the thermodynamic theory of fluid interfaces: Infinite intervals, equilibrium solutions, and minimizers, J. of Colloid and Interface Sci., 111, No. 1, 1986, pp. 119–132.

13. N.D. Alikakos, P.W. Bates, On the singular limit in a phase-field model of a phase transition, Ann. Inst. H. Poincaré, 5, 1988, pp. 1–38.

14. N.D. Alikakos, L. Bronsard, G. Fusco, Slow motion for the Cahn-Hilliard equarion via energy and spectral estimates, (In preparation), 1994.

15. N.D. Alikakos, Xinfu Chen, G. Fusco, The dynamics of an infinitesimal drop of the Cahn-Hilliard model, (In preparation), 1994.

16. N.D. Alikakos, G. Fusco, Some aspects of the dynamics of the Cahn-Hilliard equation, Resenhas Inst. Mat. Estat. Univ Sao Paulo, 1, No. 4, 1994, pp. 517–530.

17. Nicholas Alikakos, Peter W. Bates, Xinfu Chen, Convergence of the Cahn-Hilliard equation to the Hele-Show model, (To appear).

18. Nicholas Alikakos, Peter W. Bates, Giorgio Fusco, Slow motion for the Cahn-Hilliard equation in one space dimension, J. Differential Equations, 90, No. 1, 1991, pp. 81–135.

19. S. Allen, J.W. Cahn, A microscopic theory for antiphase boundary motion and its application to antiphase domain coarsening, Acta Metall., 27, 1979, pp. 1084–1095.

20. Hans Wilhelm Alt, Irena Pawlow, Existence of solutions for non-isothermal phase separation, Preprint of Inst. für Angewandte Math. der Univ. Bonn, No. 159, 1991 (To appear in European J. Appl. Math.).

21. Hans Wilhelm Alt, Irena Pawlow, A mathematical model of dynamics of non-isothermal phase separation, Preprint of Inst. für Angewandte Math. der Univ. Bonn, No. 158, 1991 (To appear in Physica $\mathbb{D}$).

22. S.A. Arsenjev, A.Ju. Gubar', D.N. Rykunov, N.K. Shelkovnikov, New type of long waves instability on shallow water: nonlinear viscosity of changing sign in tides, Dokl. USSR Acad. sci. (Доклады Акад. наук СССР), 323, No. 1, 1992, pp. 58–61 (in Russian).

23. J.M. Ball, P.J. Holmes, R.D. James, R.L. Pego, P.J. Swart, On the dynamics of fine structure, J. Nonlinear Sci., 1, No. 1, 1991, pp. 19–70.

24. G.I. Barenblatt, M. Bertsch, R. Dal Passo, V.M. Prostokishin, M. Ughi, A mathematical model of turbulent heat and mass transfer in stably stratified shear flow, J. Fluid Mech, 253, 1993, pp. 341–358.

25. G.I. Barenblatt, M. Bertsch, R. Dal Passo, M. Ughi, A degenerate pseudoparabolic regularization of a nonlinear forward-backward heat equation arising in the theory of heat and mass exchange in stably stratified turbulent shear flow, SIAM J. Math. Anal., 24, No. 6, 1993, pp. 1414–1439.

26. Peter W. Bates, Paul C. Fife, Spectral comparison principle for the Cahn-Hilliard and phase-field equations, and time scales for coarsening, Physica $\mathbb{D}$, 43, 1990, pp. 335–348.

27. Peter W. Bates, Paul C. Fife, The dynamics of nucleation for the Cahn-Hilliard equation, SIAM J. Appl. Math., 53, No. 3, 1993.

28. Peter W. Bates, Kening Lu, A Hartman-Grobman theorem for Cahn-Hilliard equations and phase-field equations, J. Dyn. Diff. Equat. (To appear).

29. Peter W. Bates, Ping J. Xun, Metastable patterns for the Cahn-Hilliard equation: Part I, J. Differential Equations (To appear).

30. Peter W. Bates, Ping J. Xun, Metastable patterns for the Cahn-Hilliard equation: Part II, Layer dynamics and slow invariant manifold, J. Differential Equations (To appear).

31. P.W. Bates, Songmu Zheng, Inertial manifolds and inertial sets for the phase-field equations, J. Dynamics Differential Equations, 4, No. 2, 1992, pp. 375–398.

32. I.M. Belkin, G.V. Vinogradov, A.I. Leonov, Rotational Instruments: A Viscosity and Physical-Mechanical Characteristic of Materials Measurements, Nauka, Moscow, 1968 (in Russian).

33. E Benoit, J.-L. Callot, F. Diener, M. Diener, Chasse au canard, Coll. Math., 31, 1980.

34. Ju.A. Berezin, G.I. Dudnikova, V.A. Novikov, N.N. Yanenko, Analytical and numerical studies of equations with coefficient at the leading derivative of variable sign, Preprint, Inst. Pure Appl. Mech., Siberian Division of USSR Acad. Sci., No. 15, 1978 (in Russian).

35. L. Bers, F. John, M. Schechter, Partial Differential Equations, volume 3A of Lectures in Appl. Math. Amer. Math. Soc., Providence, Rhode Island, seventh edition, 1990.

36. N. Bourbaki, Intégration, volume VI of Éléments de Mathématique, Hermann & Cie, Paris.

37. Lia Bronsard, Danielle Hilhorst, On the slow dynamic for the Cahn-Hilliard equation in one space dimension, Proc. Royal Soc. London, A 439, 1992, pp. 669–682.

38. J.F. Browley, C.M. Elliott, The Cahn-Hilliard gradient theiry for phase separation with non-smooth energy, Part II: numerical analysis. Euro. J. Appl. Math., 3, 1992, pp. 147–179.

39. G. Caginalp, Surface tension and supercooling in solidification theory, Lecture Notes in Physics, 216, 1984, pp. 216–226.

40. G. Caginalp, An analysis of a phase filed model of a free boundary, Arch. Rational Mech. Anal., 92, 1986, pp. 205–245.

41. J.W. Cahn, J.E. Hilliard, Free energy of a nonuniform system. I. Interfacial free energy, J. Chem. Physics, 28, 1958, pp. 258–267.

42. Jack Carr, Morton E. Gurtin, Marshall Slemrod, Structured phase transitions on a finite interval, Archive Ration. Mech. Anal., 86, No. 4, 1984, pp. 317–351.

43. A.V. Chechkin, A.V. Tur, V.V. Yanovsky, Negative viscosity and generation of dissipative solitons and zonal dissipative structures by drift waves, Preprint, Kharkov Inst. of Physics and Technology, 1991.

44. Xinfu Chen, Spectrum for the Allen-Cahn, Cahn-Hilliard, and phase-field equations for generic interfaces, Commun. in Partial Differential Equation, 9, No. 7–8, 1994, pp. 1371–1396.

45. Donald S. Cohen, James D. Murray, A generalized diffusion model for growth and dispersal in a population, J. Math. Biology, 12, 1981, pp. 237–249.

46. M.I.M. Copetti, C.M. Elliott, Kinetics of phase decomposition processes: numerical solutions to Cahn-Hilliard equation, Materials Sci. and Technology, 6, No. 3, 1990, pp. 273–283.

47. B.J. Daly, The stability properties of coupled pair of nonlinear partial difference equations, Math. Comp., 84, No. 17, 1963, pp. 346–360.

48. V.G. Danilov, G.A. Omel'yanov, E.V. Radkevich, Asymptotic solution ofa phase field system and the modified Stefan problem, Differential Equations, 31, No. 3, 1995.

49. Martin Davis, Applied Nonstandard Analysis, John Wiley & Sons, New York e.a., 1977.

50. S.Kh. Dgumagazieva, Numerical study of a partial differential equation, J. of Numer. Math. and Math. Phys., 23, No. 4, 1983, pp. 839–847 (in Russian).

51. R.J. DiPierna, Measure-valued solutions to conservation laws, Archive Ration. Mech. Anal., 88, No. 3, 1985.

52. Charles M. Elliott, The Stefan problem with a non-monotone constitutive relation, IMA J. Appl. Math., 35, 1985, pp. 257–264.

53. Charles M. Elliott, Donald A. French, Numerical studies of the Cahn-Hilliard equation for the phase separation, IMA J. Appl. Math., 38, 1987, pp. 97–128.

54. Charles M. Elliott, Donald A. French, A nonconforming finite-element method for the two-domensional Cahn-Hilliard equation, SIAM J. Numer. Anal., 26, No. 4, 1989, pp. 884–903.

55. Charles M. Elliott, Stig Larsson, Error estimates with smooth and nonsmooth data for a finite element method for the Cahn-Hilliard equation, Math. Comp., 58, No. 198, 1992, pp. 603–630.

56. Charles M. Elliott, D.A. Luckhaus, A generalized diffusion equation for phase separation of a multi-component mixture with interfacial free energy, preprint.

57. Charles M. Elliott, Andro Mikelić, Existence for the Cahn-Hilliard phase separation model with a nondifferential energy, Ann. di Mat. pura ed appl., CLVIII, N. IV, 1991, pp. 181–203.

58. Charles M. Elliott, Songmu Zheng, On the Cahn-Hilliard equation, Archive Ration. Mech. Anal., 96, No. 4, 1986, pp. 339–357.

59. C.M. Elliott, The Cahn-Hilliard model for the kinetics of phase separation, In Math. Models for Phase Change Problems. (J.E. Rodrigues, editor), number 88 in Int. Series of Numer. Math., Basel, Birkhäuser Verlag, 1989, pp. 35–73.

60. C.M. Elliott, D.A. French, F.A. Milner, A second order splitting method for the Cahn-Hilliard equation, Numer. Math., 54, 1989, pp. 575–590.

61. D.J. Eyre, Systems of Cahn-Hilliard equations, SIAM J. Appl. Math. (To appear).

62. P.C. Fife, Models for phase separation and their mathematics, In Nonlinear Partial Differential Equations and Applications (M. Mimura and T. Nishida, editors), Tokyo, Kinokuniya Pubs, 1995. (To appear).

63. A.F. Filippov, Differential Equations with Discontinuous Right-Hand Side, Nauka, Moscow, 1985 (in Russian).

64. G. Fix, Phase field methods for free boundary problems, In Free boundary problems: theory and applications (B. Fasano and M. Primicerio, editors), Pitman, 1983, pp. 580–569.

65. Daisuke Furihata, Tomohiro Onda, Masatake Mori, A numerical analysis of some phase separation problem, In Proc. of the First China-Japan Sem. om Numer. Math. (Zhong-Ci Shi and Terio Ushijima, editors), Singapure, New Jersey, London, Hong Kong, World Scientific, August 24–29, 1992, pp. 29–44.

66. Daisuke Furihata, Tomohiro Onda, Masatake Mori, A finite difference scheme for the Cahn-Hilliard equation based on a Lyapunov functional, In Nonlinear Math. Problems in Industry, number 2 in GAKUTO Int. Ser, Math. Sci. and Appl., 1993, pp. 347–358.

67. Christopher P. Grant, Slow motion on one-dimensional Cahn-Morral systems, Preprint.

68. Christopher P. Grant, Spinodal decomposition for the Cahn-Hilliard equation, Commun. in Partial Differential Equation, 18, No. 3–4, 1993, pp. 453–490.

69. Vladimir N. Grebenev, On an interfacial phenomenon for one-dimensional equation of forward-backward parabolic type (To appear).

70. V.N. Grebenev, On a solvability of non-standard problems for nonlinear parabolic equations with variable time direction, PhD dissertation, Novosibirsk State Univ., Novosibirsk, 1987.

71. J.D. Gunton, M. Droz, Introduction to the Theory of Metastable and Unstable States, volume 183 of Lect. Notes in Physics, Springer Verlag, 1983.

72. M.E. Gurtin, Thermodynamics of evolving phase boundaries in the plane, Oxford Univ. Press, 1993.

73. Morton E. Gurtin, Some results and conjectures in the gradient theory of phase transitions, In Metastability and Incompletely Posed Problems (S. Antman, J.L. Ericksen, D. Kinderlehrer, and I. Müller, editors), number 3 in The IMA Volumes in Math. and its Appl., N.Y. e.a., Springer Verlag, 1987, pp. 135–146.

74. Morton E. Gurtin, Hiroshi Matano, On the structure of equilibrium phase transitions within the gradient theory of fluids, Quart. Appl. Math., XLVI, No. 2, 1988, pp. 301–317.

75. Philip Hartman, Ordinary Differential Equations, John Wiley & Sons, New York, e.a., 1964.

76. Michael Hazewinkel, Experimental mathematics, In Math. and Comput. Sci., Proc. CWI Symp., Amsterdam, 25 November, 1983, Amsterdam, e.a., 1986, pp. 193–234.

77. Michael Hazewinkel, Johan F. Kaashoek, Bart Leynse, Pattern formation for a one dimensional evolution equation based on Thom's river basin model, Technical Report 8519/B, Econometric Inst., Erasmus Univ., Rotterdam, 1985.

78. Michael Hazewinkel, Johan F. Kaashoek, Bart Leynse, Liesegang rings, pattern formation at phase transition: experiments and mathematical modelling, Technical report, Econometric Inst., Erasmus Univ., Rotterdam, 1987.

79. Dan Henry, Geometric Theory of Semilinear parabolic Equations, volume 840 of Lect. Notes Math., Springer Verlag, Berlin, e.a., 1981.

80. Klaus Höllig, Existence of infinitely many solutions for a forward backward heat equation, Trans. Amer. Math. Soc., 278, No. 1, July 1983, pp. 299–316.

81. Klaus Höllig, John A. Nohel, Diffusion equation with a nonmonotone constitutive function, In Systems of Nonlinear Partial Different. Equat.: Proc NATO/London Math. Soc. Conf., Oxford, 1982 (J.M. Ball, editor), pp. 409–422. D. Reidel Publishing Company, 1983.

82. J. Hunter, M. Slemrod, Viscoelastic fluid flow exhibiting hysteresis phase changes, Phys. Fluids, 26, 1983, pp. 2345–2351.

83. M. Johnson, D. Segalman, A model for viscoelastic fluid behavior which allows non-affine deformation, J. Non-Newtonian Fluid Mech., 2, No. 3, 1977, pp. 255–270.

84. S.L. Kamenomostskaja, On Stefan problem, Matem. Sbornik (Матем. сборник), 53(95), No. 4, 1961, pp. 489–514 (in Russian).

85. V.F. Kim, G.I. Zelinskaja, Numerical solving of equations with viscosity coefficient of variable sign, Numer. Methods in Mech. Continuum Media (Численные методы в механике сплошной среды), Novosibirsk, 12, No. 1, 1981, pp. 54–68 (in Russian).

86. David Kinderlehrer, Pablo Pedregal, Weak convergence of integrands and the Young measure representation, SIAM J. Math. Anal., 23, No. 1, January 1992, pp. 1–19.

87. E.Ja. Kogan, N.E. Molevich, Collapse of acoustic waves in non-equilibrium molecular gas, J. Techn. Phys. (Техническая физика), 56, No. 5, 1986, pp. 941–943 (in Russian).

88. E.Ja. Kogan, N.E. Molevich, A.N. Orajevskii, Structure of nonlinear acoustic waves in non-equilibrium oscillatory excitatory gas, J. Techn. Phys. Letters (Техническая физика), 13, No. 14, 1987, pp. 836–838 (in Russian).

89. Robert V. Kohn, Peter Sternberg, Local minimizers and singular perturbations, Proc. Royal Soc. Edinburgh, A 111, 1989, pp. 69–84.

90. Ju.P. Krasovskii, On development of a layer structure of the thermocline, Sea Geophys. J. (Геофизика моря), No. 5, 1986, pp. 18–22 (in Russian).

91. A.A. Lacey, M. Shillor, The existence and stability of regions with superheating in the classical two-phase one-dimensional Stefan problem with heat sources, IMA J. Appl. Math., 30, 1983, pp. 215–230.

92. O.A. Ladyženskaja, V.A. Solonnikov, N.N. Ural'ceva, Linear and Quasilinear Equations of Parabolic Type, volume 23 of Trans. Math. Monographs, Amer. Math. Soc., Providence, RI, 2nd edition, 1988.

93. Alan V. Lair, Uniqueness for a forward backward diffusion equation, Trans. Amer. Math. Soc., 291, No. 1, September 1985, pp. 311–317.

94. Alan V. Lair, Uniqueness for a forward backward diffusion equation with smooth constitutive function, Applicable Anal., 29, 1988, pp. 177–189.

95. N.A. Lar'kin, V.A. Novikov, N.N. Yanenko, Nonlinear Equations of Variable Type, Nauka, Novosibirsk, 1983 (in Russian).

96. M.M. Lavrent'ev Jr., Strong a priori estimates of solutions of equations of variable type, Siberian J. Diff. Equations (To appear).

97. M.M. Lavrent'ev Jr., Estimates for solutions of a variable type equation, Math. modelling (Математическое моделирование), 1, No. 11, 1989, pp. 132–138 (in Russian).

98. M.M. Lavrent'ev Jr., A priory smoothness of solutions for number of equations of changing type. Math. modelling (Математическое моделирование), 2, No. 9, 1990, pp. 145–153 (in Russian).

99. J.L. Lions, Quelques méthodes de résolution des problèmes aux limites non linéaires, Dunod Gauthier-Villars, Paris, 1969.

100. N.N. Maslova, Numerical analysis of stability of self-similar solutions for equations with viscosity coefficient of variable sign, PhD thesis, Novosibirsk, 1985.

101. N.N. Maslova, V.A. Novikov, N.N. Yanenko, Some selfsimilar solutions for a class of hydrodynamic equations with viscosity of variable sign, Soviet Math. Dokl., 30, No. 1, 1984, pp. 87–90.

102. Luciano Modica, The gradient theory of phase transitions and the minimal interface criterion, Archive Ration. Mech. Anal., 98, No. 2, 1987, pp. 123–142.

103. A.S. Monin, A.V Yaglom, Statistical Fluid Mechanics, volume 1, MIT Press, Cambridge, 1971.

104. J.E. Morral, J.W. Cahn, Spinodal decomposition in ternary systems, Acta Metall., 19, 1971, pp. 1037–1045.

105. J.D. Murray, A pre-pattern formation mechanism for animal coat markings, J. Theoret. Biology, 88, 1981, pp. 161–199.

106. E.B. Nauman, N.P. Balsara, Spatially local minimizers of the Landau-Ginsburg functional, Quart. Appl. Math., XLVI, No. 2, 1988, pp. 375–379.

107. B. Nicolaenko, B. Scheurer, R. Temam, Some global dynamical properties of a class of pattern formation equations, Comm. in Partial Different. Equat., 14, No. 2, 1989, pp. 245–297.

108. John A. Nohel, Robert L. Pego, Athanasios Tzavaras, Stability of discontinuous steady states in shearing motions of a non-Newtonian fluid, Proc. Royal Soc. Edinburgh, $\mathbb{A}$ 115, No. 1, 1990, pp. 39–59.

109. A. Novick-Cohen, R.L. Pego, Stable patterns in a viscous diffusion equation, Trans. Amer. Math. Soc., 324, No. 1, 1991, pp. 331–351.

110. Amy Novick-Cohen, Lee A. Segel, Nonlinear aspects of the Cahn-Hilliard equation, Physica $\mathbb{D}$, 10, 1984, pp. 277–298.

111. V.A. Novikov, Existence and non-existence theorems for a class of equations of variable type, Dokl. USSR Acad. sci. (Доклады Академии Наук СССР), 253, No. 6, 1980, pp. 1311–1313 (in Russian).

112. V.A. Novikov, Qualitative properties of solutions of a class of equations of variable type, In Numer. meth. for balance equation, Berlin, 1980, pp. 64–73.

113. V.A. Novikov, I.V. Shvab, On non-uniqueness and "stability" of some solutions of nonlinear forward backward parabolic equations, Numer. Methods in Mech. Continuum Media (Численные методы в механике сплошной среды), Novosibirsk, 16, No. 4, 1985, pp. 53–76 (in Russian).

114. J. Oldroyd, Non-Newtonian effects in steady motion of some idealized elastico-viscous liquids, Proc. Roy. Soc. London, A 245, No. 1240, 1958, pp. 278–297.

115. G.A. Omel'yanov, E.V. Radkevich, V.G. Danilov, Justification of the phase field system asymptotic solution and modified Stefan problem, Matem. Sbornik (Математический сборник), 186, No. 12, 1995, pp. 63–80 (in Russian).

116. N.C. Owen, J. Rubinstein, P. Sternberg, Minimizers and gradient flows for singularly perturbed bi-stable potentials with dirichlet condition, Proc. Royal Soc. London, $\mathbb{A}$ 429, 1990, pp. 505–532.

117. R.L. Pego, Front migration in the nonlinear Cahn-Hilliard equation, Proc. Royal Soc. London, $\mathbb{A}$ 422, 1989, pp. 261–278.

118. O. Penrose, P. Fife, Thermodynamically consistent model of phase-field type for kinetics of phase transition, Physica $\mathbb{D}$, 43, 1990, pp. 44–62.

119. S.G. Pjatkov, Solvability of initial boundary value problems for a nonlinear forward backward parabolic equation, Preprint, Inst. Math., Siberian Division of USSR Acad. Sci., No. 16, 1987 (in Russian).

120. S.G. Pjatkov, A.G. Podgaev, On solvability of a boundary value problem for a nonlinear parabolic equation with variable time direction, Siberian Math. J. (Сибирский математический журнал), 28, No. 3, 1987, pp. 184–192 (in Russian).

121. P.I. Plotnikov, Forward backward parabolic equations and hysteresis effect, Dokl. Russian Acad. sci. (Доклады Российской Академии Наук), 330, No. 6, 1993, pp. 691–693 (in Russian).

122. P.I. Plotnikov, Passing to the limit with respect to viscosity in a forward backward parabolic equation, Differents. Equat. (Дифференциальные уравнения), 30, No. 4, 1994, pp. 665–674 (in Russian).

123. P.I. Plotnikov, Forward-backward parabolic equations and hysteresis, (In preparation), 1996.

124. P.I. Plotnikov, V.N. Starovoitov. Stefan problem as the limit of the phase-field system, Differential Equations, 29, No. 3, 1993.

125. A.G. Podgaev, On a boundary value problem for Korteweg-de Vries-Burger equation with diffusion coefficient of variable sign, In Non-classical Equations of Mathematical Physics (Неклассические уравнения математической физики) (V.N. Vragov, editor), Novosibirsk, Inst. of Math. USSR Academy of Sciences, Siberian Division, 1986, pp. 97–107 (in Russian).

126. E.V. Radkevich, About asymptotic solution of a phase field system, Differential Equations, 29, No. 3, 1993.

127. S.M. Rankin, III, Semilinear evolution equations in Banach spaces with applications to parabolic partial differential equations, Trans. Amer. Math. Soc. (To appear).

128. Abraham Robinson, Non-Standard Analysis, Studies in Logic and Foundations of Math. North-Holland, 1966.

129. B.N. Sadovskii, On uniqueness conditions for ordinary differential equations, Uspechi Matem. Nauk (Успехи математических наук), 21, No. 5 (131), 1966, pp. 263–265 (in Russian).

130. Weixi Shen, Songmu Zheng, On the coupled Cahn-Hilliard equations, Commun. in Partial Differential Equation, 18, No. 3&4, 1993, pp. 701–727.

131. Ju.I. Shokin, N.N. Yanenko, The First Differential Approximation Method. Applications to Gas Dynamics, Nauka, Novosibirsk, 1985 (in Russian).

132. B.Ju. Skobelev, N.N. Yanenko, Stability of stationary flows in a fluid with viscosity of variable sign, Numer. Methods in Mech. Continuum Media (Численные методы в механике сплошной среды), Novosibirsk, 13, No. 5, 1982, pp. 61–72 (in Russian).

133. V.P. Skripov, A.V.Skripov, Spinodal decomposition (phase transition via unstable states), Sov. Phys. Usp., 22, 1979, pp. 389–410.

134. M. Slemrod, Measure-valued solutions to a backward-forward heat equation: a conference report, In Nonlinear Evolution Equations that Change Type (Barbara Lee Keyfitz and Michael Sherer, editors), number 27 in The IMA Volumes in Math. and Appl., N.Y., e.a., Springer Verlag, 1990, pp. 232–242.

135. M. Slemrod, Dynamics of measure valued solutions to a backward forward heat equation, J. Dyn. Diff. Equat., 3, No. 1, 1991, pp. 1–28.

136. H.M. Soner, Convergence of the phase-field equations to the Mullins-Sekerka problem with kinetic undercooling, Archive Rational Mech. Anal., 131, No. 2, 1995, pp. 139–197.

137. Peter Sternberg, The effect of a singular perturbation on nonconvex variational problems, Archive Ration. Mech. Anal., 101, No. 3, 1988, pp. 109–260.

138. Gilbert Strang, M. Abdel-Naby, Notes on softening and local instability, Number 3 in Lect. Notes in Engineering. Springer Verlag, 1983.

139. L. Tartar, Compensated compactness and applications to partial differential equations, In Nonlinear Anal. and Mech. Heriott-Watt Symp. IV, Research Notes in Math., London, Pitman, 1979, pp. 136–192.

140. R. Thom, Symmetries gained and lost, In Mathematical Physics and Physical Mathematics (K. Maurin and R. Raczka, editors), Dordrecht, Reidel, 1976.

141. J.D. van der Waals, The thermodynamic theory of capillarity flow under the hypothesis of a continuous variation of density, Verhandel. Konink. Akad. Weten. Amsterdam (Sec 1), 1, No. 8, 1983. English transl. by Rowlinson J.S. see in J. Stat. Phys., 20, 1979, pp. 197–244.

142. V.P. Voronko, Difference schemes for nonlinear equations of variable type, Numer. Methods in Mech. Continuum Media (Численные методы в механике сплошной среды), Novosibirsk, 8, No. 5, 1977, pp. 48–52 (in Russian).

143. N.N. Yanenko, G.P. Kurbatkin, V.N. Krupchatnikov, M.Sh. Eicher, On a model of atmosphere circulation with local turbulence coefficient of variable sign, Numer. Methods in Mech. Continuum Media (Численные методы в механике сплошной среды), Novosibirsk, 7, No. 1, 1976, pp. 137–153 (in Russian).

144. N.N. Yanenko, V.A. Novikov, On a model of a fluid with viscosity coefficient of variable sign, Numer. Methods in Mech. Continuum Media

(Численные методы в механике сплошной среды), Novosibirsk, 4, No. 2, 1973, pp. 142–147 (in Russian).

145. N.N. Yanenko, A.S. Solovjev, On self-oscillations in one dimensional wave systems, Numer. Methods in Mech. Continuum Media (Чиrленные методы в механике сплошной среды), Novosibirsk, 11, No. 7, 1980, pp. 128–140 (in Russian).

146. Jingxue Yin, On the existence of nonnegative continuous solutions of the Cahn-Hilliard equation, J. Differential Equations, 97, No. 2, 1992, pp. 310–327.

147. Jiongmin Yong, Songmu Zheng, Feedback stabilization and optimal control for the Cahn-Hilliard equation, Nonlinear Anal.: Theory, Methods and Appl., 17, No. 5, 1991, pp. 431–444.

148. T.I. Zelenjak, On an equation with diffusion coefficient of variable sign, In Math. Problems of Chemistry (Математические проблемы химии), Novosibirsk, No. 1, 1975, pp. 111–115 (in Russian).

149. T.I. Zelenjak, Qualitative properties of a solutions of semilinear mixed boundary value problems for equations of parabolic type, Matem. Sbornik (Математический сборник), 107, No. 3(146), 1977, pp. 486–510 (in Russian).

150. T.I. Zelenjak, V.A. Novikov, N.N. Yanenko, On properties of solutions of nonlinear equations of variable type, Numer. Methods in Mech. Continuum Media (Численные методы в механике сплошной среды), Novosibirsk, 5, No. 4, 1974, pp. 35–47 (in Russian).

151. Songmu Zheng, Asymptotic behavior of the solution to the Cahn-Hilliard equation, Applicable Anal., 23, 1986, pp. 165–184.

152. A. Zygmund, Trigonometrical Series, Monografje Matematiczne, Warszawa, 1935.

Translated by the author

Chapter 2

On solvability of mixed problem for hyperbolic equation in a quadrant

V.A.Dobrushkin[1] and T.R.Jum[2]

Abstract. Conditions of normal solvability of an initial-boundary value problem for the two dimnesional wave equation in a quadrant on the plane was found in the paper [1]. Present work is dedicated to the investigation of the hyperbolic equation

$$\left(\frac{\partial^2}{\partial t^2} - \frac{\partial^2}{\partial x^2} - \frac{\partial^2}{\partial x^2} - p_0\frac{\partial}{\partial t} - p_0\frac{\partial}{\partial x} - p_0\frac{\partial}{\partial y} - p_3\right)u(x,y,t) = 0$$

with constant coefficients p_j $(j = 1,2,3)$ in the domain $0 < x, y < \infty$, given the boundary conditions which contain the common differential operator of the firs order. Conditions on the coefficients which ensure the normal solvability of the considered problem are found.

1. Formulation of problem

In the domain $\{(x,y,t) \in R^3 : x > 0, y > 0, t > 0\}$ we consider the equation

$$u_{tt} - u_{xx} - u_{yy} - p_0u_t - p_1u_x - p_2u_y - p_3u = 0; \tag{1.1}$$

by u_{tt}, u_{xx}, u_{yy}, u_t, u_x, u_y we denote the first and second partial derivatives of the function $u(x,y,t)$. We suppose that on the boundary the following boundary conditions are given:

$$(u_t - ay_x - bu_y + cu)\big|_{x=0} = f_1(y,t), \tag{1.2}$$

$$(u_t - \alpha y_x - \beta u_y + \gamma u)\big|_{y=0} = f_2(x,t), \tag{1.3}$$

where $a, b, c, \alpha, \beta, \gamma, p_0, p_1, p_2, p_3$ are some real constants. The initial data are supposed to be homogeneous

$$u(x,y,0) = 0, \quad u_t(x,y,0) = 0, \tag{1.4}$$

at the point $x = 0$, $y = 0$ the condition on the edge is given [2]:

$$u(0,0,t)$$

[1]Institute of Mathematics, Academy of Sciences of Belorussia
[2]Institute of Mathematics, Academy of Sciences of Belorussia

We define the function $w(x,y,t)$ by the equality

$$w(x,y,t) = u(x,y,t)\exp\left\{-\frac{1}{2}(p_0 t - p_1 x - p_2 y)\right\}. \tag{1.6}$$

Then it satisfies the wave equation:

$$w_{tt} - w_{xx} - w_{yy} + pw = 0, \tag{1.7}$$

the following boundary conditions

$$[w_t - aw_x - bw_y + dw]\big|_{x=0} = q_1(y,t), \quad (y,t) \in R^2_+, \tag{1.8}$$

$$[w_t - \alpha w_x - \beta w_y + hw]\big|_{y=0} = q_2(x,t), \quad (x,t) \in R^2_+, \tag{1.9}$$

and the initial data

$$w(x,y,0) = 0, \quad w_t(x,y,0) = 0 \tag{1.10}$$

and the condition on the edge

$$w(0,0,t) = 0. \tag{1.11}$$

Here

$$p = \frac{1}{4}(p_0^2 - p_1^2 - p_2^2) - p_3, \tag{1.12}$$

$$d = \frac{1}{2}(p_0 + ap_1 + bp_2) + c, \tag{1.13}$$

$$h = \frac{1}{2}(p_0 + \beta p_1 + \alpha p_2) + \gamma, \tag{1.14}$$

$$q_1(y,t) = \exp\left\{-\frac{1}{2}(p_0 t - p_2 y)\right\} f_1(y,t), \tag{1.15}$$

$$q_2(x,t) = \exp\left\{-\frac{1}{2}(p_0 t - p_1 x)\right\} f_2(x,t). \tag{1.16}$$

By $\Pi(r,m)$ we denote the class of functions $f(\mathbf{x},t)$ ($\mathbf{x} \in D \subset R^n$, $n > 0$) which have the integrable in square Laplace transform with respect to t:

$$f^{\mathbf{L}}(\mathbf{x},\lambda) = \int_0^\infty f(\mathbf{x},t),$$

it is regular with respect to λ in the right half-plane $\{\lambda \in \mathbf{C} : \mathrm{Re}\lambda > r > 0\}$ and satisfies the estimate on infinity

$$|f^{\mathbf{L}}(\mathbf{x},\lambda)| < \mathrm{const}|\lambda|^m.$$

Let $w \in \Pi(r, m)$. Applying the integral Laplace transform with respect to time variable t to problem (1.7)–(1.11), we come to

$$(\lambda^2 + p)v - \Delta v = 0, \tag{1.17}$$

$$[(\lambda + d)v - av_x - bv_y]\big|_{x=0} = g_1(y, \lambda), \tag{1.18}$$

$$[(\lambda + h)v - \beta v_x - \alpha v_y]\big|_{y=0} = g_2(x, \lambda), \tag{1.19}$$

where Δ is the Laplace operator,

$$v(x, y, \lambda) = w^L(x, y, \lambda) = \int_0^\infty w(x, y, t)e^{-\lambda t}, \tag{1.21}$$

$$g_j(x, y, \lambda) = q_j^L(x, y, \lambda) = \int_0^\infty q_j(x, y, t)e^{-\lambda t} \quad (j = 1, 2). \tag{1.22}$$

2. Construction of formal solution to problem (1.17)–(1.20)

Following [1], we introduce operators of projection on one of the quadrant sides:

$$B_{\mathbf{A}}^x v = [(\lambda + d)v - bv_y]\big|_{x=0}, \tag{2.1}$$

$$B_{\mathbf{B}}^x v = -av_x\big|_{x=0}, \tag{2.2}$$

$$B_{\mathbf{A}}^y v = [(\lambda + h)v - \beta v_x]\big|_{y=0}, \tag{2.3}$$

$$B_{\mathbf{B}}^y v = -\alpha v_y\big|_{y=0}, \tag{2.4}$$

The solution to problem (1.17)–(1.20) is sought in the form [2]

$$v = V_{\mathbf{A}}^x\varphi + V_{\mathbf{B}}^x\varphi + V_{\mathbf{A}}^y\psi + V_{\mathbf{B}}^y\psi, \tag{2.5}$$

where $\varphi(y, \lambda)$, $\psi(y, \lambda)$ are unknown as yet functions, and the operators V_A^x, V_B^x, V_A^y, V_B^y are defined by the equalities:

$$V_{\mathbf{A}}^x\varphi = \frac{1}{\pi}\int_0^\infty \frac{x\sqrt{\lambda^2+p}}{\sqrt{x^2+(y-\xi)^2}} K_1\big(\sqrt{\lambda^2+p}\sqrt{x^2+(y-\xi)^2}\big)\Phi(\xi, \lambda)d\xi, \tag{2.6}$$

$$V_{\mathbf{B}}^x\varphi = \frac{1}{\pi a}\int_0^\infty K_0\big(\sqrt{\lambda^2+p}\sqrt{x^2+(y-\xi)^2}\big)\varphi(\xi, \lambda)d\xi, \tag{2.7}$$

$$V_{\mathbf{A}}^y\psi = \frac{1}{\pi}\int_0^\infty \frac{y\sqrt{\lambda^2+p}}{\sqrt{y^2+(y-\eta)^2}} K_1\big(\sqrt{\lambda^2+p}\sqrt{y^2+(x-\eta)^2}\big)\Psi(\eta, \lambda)d\eta, \tag{2.8}$$

$$V_{\mathbf{B}}^{y}\varphi = \frac{1}{\pi\alpha}\int_{0}^{\infty} K_0\left(\sqrt{\lambda^2+p}\sqrt{y^2+(x-\eta)^2}\right)\psi(\eta,\lambda)d\eta. \tag{2.9}$$

Here

$$\Phi(y,\lambda) = -\frac{1}{b}\int_{0}^{y}\varphi(\xi,\lambda)\exp\left\{\frac{\lambda+d}{b}(y-\xi)\right\}d\xi, \tag{2.10}$$

$$\Psi(\eta,\lambda) = -\frac{1}{\beta}\int_{0}^{\eta}\psi(\theta,\lambda)\exp\left\{\frac{\lambda+h}{\beta}(\eta-\theta)\right\}d\theta, \tag{2.11}$$

K_0, K_1 are McDonald functions, branches of the radicals $\sqrt{\lambda^2+p}$ and $\sqrt{y^2+(x-\eta)^2}$, $\sqrt{x^2+(y-\xi)^2}$ take positive values as their arguments are real and positive.

Function (2.5) automatically satisfies eqaution (1.17). By virtue of validity of equalities

$$B_{\mathbf{A}}^{x}V_{\mathbf{A}}^{x} = I, \quad B_{\mathbf{B}}^{x}V_{\mathbf{B}}^{x} = I, \quad B_{\mathbf{A}}^{y}V_{\mathbf{A}}^{y} = I, \quad B_{\mathbf{B}}^{y}V_{\mathbf{B}}^{y} = I, \tag{2.12}$$

on the operators (2.1)–(2.4) and (2.6)–(2.9), we obtain an equations system to determine the functions φ and ψ by substitution of (2.5) the assumed form of solution into boundary conditions (1.18), (1.19):

$$(2I + A)\vec{\Phi} = \vec{g}, \tag{2.13}$$

where I is the identity operator, $\vec{\Phi}^T = (\varphi(y,\lambda), \psi(x,\lambda))$, $\vec{g}^T(g_1(y,\lambda), g_2(y,\lambda))$, "T" stands for transposition, and elemnets of the 2×2 operator matrix A can be found from formulae:

$$\begin{aligned} a_{11} &= B_{\mathbf{A}}^{x}V_{\mathbf{B}}^{x} + B_{\mathbf{B}}^{x}V_{\mathbf{A}}^{x}, \\ a_{12} &= (B_{\mathbf{A}}^{x} + B_{\mathbf{B}}^{x})(V_{\mathbf{A}}^{y} + V_{\mathbf{B}}^{y}), \\ a_{21} &= (B_{\mathbf{A}}^{y} + B_{\mathbf{B}}^{y})(V_{\mathbf{A}}^{x} + V_{\mathbf{B}}^{x}), \\ a_{22} &= B_{\mathbf{A}}^{y}V_{\mathbf{B}}^{y} + B_{\mathbf{B}}^{y}V_{\mathbf{A}}^{y}. \end{aligned} \tag{2.14}$$

3. Condition on convergence of integrals (2.6)–(2.9)

We will show that integrals (2.6)–(2.9) exist if simulteneously one of conditions

$$\begin{aligned} &1)\ \beta < 0, \\ &2)\ \beta > 1, \\ &3)\ \beta = 1, h < 0 \end{aligned} \tag{3.1}$$

and one of conditions

$$1a)\ b < 0,$$

$$2a)\ b > 1, \tag{3.2}$$

$$3a)\ b = 1, d < 0$$

are fulfilled. We will also find the value of the positive constant s which ensures the fulfilment $\phi(\xi, \lambda)$, $\psi(\xi, \lambda) \in \Pi(r, m)$.

Indeed, let us consider, for instance, one of integrals which enter the right hand side of inequality (2.8):

$$\int_0^\infty \frac{y\sqrt{\lambda^2 + p}}{\sqrt{y^2 + (\theta + \eta)^2}} K_1\left(\sqrt{\lambda^2 + psqrty^2 + (\theta + \eta)^2}\right) \exp\left\{\frac{\lambda + h}{\beta}\eta\right\} d\eta, \tag{3.3}$$

As $z \to \infty$ the McDonald functions $K_\nu(z) \sim \sqrt{\frac{\pi}{2z}} \exp\{-z\}$, then the convergence of integral (3.3) is provided by the inequality:

$$\frac{\mathrm{Re}\lambda + h}{\beta} - \mathrm{Re}\sqrt{\lambda^2 + p} < 0, \tag{3.4}$$

where the branch of the radical $\sqrt{\lambda^2 + p}$ is taken to satisfy $\mathrm{Re}\sqrt{\lambda^2 + p} \geq 0$ at $\lambda \geq 0$ and $p \geq 0$.

In what follows we will assume that

$$\mathrm{Re}\lambda + h > 0. \tag{3.5}$$

It is obvious that the condition 1) implies the fulfilment of inequality (3.4), and, consequently, the convergence of integral (3.3).

Now let investigate the case of positive β. By virtue of (3.5), instead of (3.4) we have

$$\beta^{-2}(\mathrm{Re}\lambda + h)^2 < \left(\mathrm{Re}\sqrt{\lambda^2 + p}\right)^2.$$

For an arbitrary complex z we have

$$\mathrm{Re}^2 z > \mathrm{Re} z^2,$$

then

$$\mathrm{Re}^2\sqrt{\lambda^2 + p} \geq \mathrm{Re}^2\lambda - \mathrm{Im}^2\lambda + p.$$

Let us consider the inequality

$$\beta^{-2}(\mathrm{Re}\lambda + h)^2 < \mathrm{Re}^2\lambda - \mathrm{Im}^2\lambda + p, \tag{3.6}$$

which desribes a set M on the complex plane λ. We will find the requirements on the coefficients for the intersection of the set M and the halfplane $\mathrm{Re}\lambda \geq 0$ contains the beam $\mathrm{Re}\lambda \geq$ const.

We have

$$(\beta^{-2} - 1)\mathrm{Re}^2\lambda + 2\beta^{-2}h\mathrm{Re}\lambda + \mathrm{Im}^2\lambda + h^2\beta^{-2} - p < 0.$$

Let $\beta^{-2}-1<0$, i.e., $\beta>1$. Then (3.6) is identical to the inequality:

$$\left(\text{Re}\lambda-\frac{h}{b^2-1}\right)^2-\frac{\text{Im}^2\lambda}{1-\beta^{-2}}-\frac{p(\beta^{-2}-1)+h^2\beta^{-2}}{(\beta^{-2}-1)^2}>0. \tag{3.7}$$

If

$$p(\beta^{-2}-1)+h^2\beta^{-2}>0, \tag{3.8}$$

then the domain M bounded by branches of the hyperbola corresponds to inequality (3.7) on the plane λ, and

$$\text{Re}\lambda\geq\frac{\beta^2\sqrt{h^2\beta^{-2}+p(\beta^{-2}-1)}+h}{\beta^2-1} \tag{3.9}$$

for all λ from M. To guarantee the fulfilment of conditions (3.5) and (3.9) the constant r which defines the space $\Pi(r,m)$ must take the values

$$r\geq\max\left\{\frac{|h|-h}{2},\frac{\beta^2\sqrt{h^2\beta^{-2}+p(\beta^{-2}-1)}+h}{\beta^2-1}\right\}.$$

If

$$p(\beta^{-2}-1)+h^2\beta^{-2}>0, \tag{3.10}$$

then the set M is placed between the branches of the hyperbola, and the whole real axis belongs to M.

The constant r is responsible for the fulfilmnet of condition (3.5) only, i.e.,

$$r\geq\frac{|h|-h}{2}.$$

If

$$p(\beta^{-2}-1)+h^2\beta^{-2}=0,$$

then the domain M is situated between two intersecting lines with the point of intersection $\left(\frac{h}{\beta^2-1},0\right)$. Then

$$r\geq\max\left\{\frac{|h|-h}{2},\frac{h}{\beta^2-1}\right\}.$$

In a similar way we will check that at $\beta=1$, $h<0$, the set of solutions M of inequality (3.6), is the domain with the boundary given by the parabola

$$2h\text{Re}\lambda+\text{Im}^2\lambda+h^2-p=0.$$

Here

$$r\geq\max\left\{\frac{|h|-h}{2},\frac{p-h}{2h}\right\}.$$

So, the conditions 2) and 3) from (3.1) guarantee the fulfilment of inequality (3.6), and, consequently, the inequality (3.4) in certain unbounded domain in the right half plane. That guarantees the convrgence of integral (3.3) at least in the same domain.

Analogously one can verify that the convergence of integral (2.6) is guaranteed by conditions (3.2).

4. Proof of the solvability of equation (2.13)

Calculations show that

$$a_{11}f(y) = \frac{\lambda+d}{\pi a}\int_0^\infty f(\xi)K_0\left(\sqrt{\lambda^2+p}|y-\xi|\right)d\xi +$$

$$+\frac{b}{\pi a}\int_0^\infty f(\xi)K_1\left(\sqrt{\lambda^2+p}|y-\xi|\right)\sqrt{\lambda^2+p}\,\mathrm{sign}(y-\xi)d\xi +$$

$$+\frac{a}{\pi b}\int_0^\infty \frac{\sqrt{\lambda^2+p}}{|y-\xi|}K_1\left(\sqrt{\lambda^2+p}|y-\xi|\right)d\xi\int_0^\xi f(\eta)\exp\left\{\frac{\lambda+d}{b}(\xi-\eta)\right\}d\eta, \quad (4.1)$$

$$a_{12}f(y) = \frac{1}{\pi\alpha}\int_0^\infty f(\eta)\Bigg[(\lambda+d)K_0\left(\sqrt{\lambda^2+p}\sqrt{y^2+\eta^2}\right)+$$

$$+\frac{\sqrt{\lambda^2+p}}{\sqrt{y^2+\eta^2}}(by-a\eta)K_1\left(\sqrt{\lambda^2+p}\sqrt{y^2+\eta^2}\right)\Bigg]d\eta -$$

$$-\frac{y}{\pi\beta}\int_0^\infty \frac{\lambda^2+p}{y^2+\eta^2}(by-a\eta)K_0\left(\sqrt{\lambda^2+p}\sqrt{y^2+\eta^2}\right)d\eta\int_0^\eta f(\theta)\exp\left\{\frac{\lambda+h}{\beta}(\eta-\theta)\right\}d\theta\cdot$$

$$-\frac{\sqrt{\lambda^2+p}}{\pi\beta}\int_0^\infty\left[\frac{(\lambda+d)y}{\sqrt{y^2+\eta^2}}-\frac{2ay\eta+b(\eta^2-y^2)}{(y^2+\eta^2)^{3/2}}\right]K_1(\sqrt{\lambda^2+p}\sqrt{y^2+\eta^2})d\eta\times$$

$$\times\int_0^\eta f(\theta)\exp\left\{\frac{\lambda+h}{\beta}(\eta-\theta)\right\}d\theta, \quad (4.2)$$

$$a_{21}f(x) = \frac{1}{\pi b}\int_0^\infty f(\xi)\Bigg[(\lambda+h)K_0\left(\sqrt{\lambda^2+p}\sqrt{x^2+\xi^2}\right)+$$

$$+\frac{\sqrt{\lambda^2+p}}{\sqrt{x^2+\xi^2}}(\beta x-\alpha\xi)K_1\left(\sqrt{\lambda^2+p}\sqrt{x^2+\xi^2}\right)\Bigg]d\xi -$$

$$-\frac{x}{\pi b}\int_0^\infty \frac{\lambda^2+p}{x^2+\xi^2}(\beta x-\alpha\xi)K_0\big(\sqrt{\lambda^2+p}\sqrt{x^2+\xi^2}\big)d\xi\int_0^\xi f(\eta)\exp\left\{\frac{\lambda+d}{b}(\xi-\eta)\right\}d\eta$$

$$-\frac{\sqrt{\lambda^2+p}}{\pi b}\int_0^\infty\left[\frac{(\lambda+h)x}{\sqrt{x^2+\xi^2}}-\frac{2ax\xi+\beta(\xi^2-x^2)}{(x^2+\xi^2)^{3/2}}\right]K_1(\sqrt{\lambda^2+p}\sqrt{x^2+\xi^2})d\xi\times$$

$$\times\int_0^\xi f(\eta)\exp\left\{\frac{\lambda+d}{b}(\xi-\eta)\right\}d\eta, \tag{4.3}$$

$$a_{22}f(x)=\frac{\lambda+h}{\pi\alpha}\int_0^\infty f(\eta)K_0\big(\sqrt{\lambda^2+p}|x-\eta|\big)d\eta+$$

$$+\frac{\beta}{\pi\alpha}\int_0^\infty f(\eta)K_1\big(\sqrt{\lambda^2+p}|x-\eta|\big)\sqrt{\lambda^2+p}\mathrm{sigr}\,(x-\eta)d\eta+$$

$$+\frac{\alpha}{\pi\beta}\int_0^\infty\frac{\sqrt{\lambda^2+p}}{|x-\eta|}K_1\big(\sqrt{\lambda^2+p}|x-\eta|\big)d\eta\int_0^\eta f(\theta)\exp\left\{\frac{\lambda+h}{\beta}(\eta-\theta)\right\}d\theta, \tag{4.4}$$

By analogy with [1] we present operators (4.1)–(4.4) in the neighborhood of zero in the form:

$$a_{ij}=a_{ij}^0+a_{ij}^R,\quad (i,j=1,2), \tag{4.5}$$

where a_{ij}^R $(i,j=1,2)$ are compact operators in $L^2(0,R)$ for an arbitrary $R>0$, and the components a_{ij}^0 $(i,j=1,2)$ will be introduced below. We replace the McDonald functions K_ν, $(\nu=0,1,\ldots)$ by their asymptotical presentations in a neighborhood of zero. Making the corresponding limit passages, we come to

$$a_{11}^0=\left(\frac{b}{a}-\frac{a}{b}\right)S, \tag{4.6}$$

$$a_{12}^0=\left(\frac{b}{\beta}-\frac{a}{\alpha}\right)S_1+\left(\frac{b}{\alpha}-\frac{a}{\beta}\right)S_2, \tag{4.7}$$

$$a_{21}^0=\left(\frac{\beta}{b}-\frac{\alpha}{a}\right)S_1+\left(\frac{\beta}{a}-\frac{\alpha}{b}\right)S_2, \tag{4.8}$$

$$a_{22}^0=\left(\frac{\beta}{\alpha}-\frac{\alpha}{\beta}\right)S, \tag{4.9}$$

where the singular operators S, S_1 and S_2 are defined by the following equalities:

$$Sf(y)=\frac{1}{\pi}\int_0^\infty\frac{f(\xi)}{y-\xi}d\xi, \tag{4.10}$$

$$S_1 f(y) = \frac{1}{\pi} \int_0^\infty f(\xi) \frac{\xi}{\xi^2 + y^2} d\xi, \tag{4.11}$$

$$S_2 f(y) = \frac{y}{\pi} \int_0^\infty \frac{f(\xi)}{\xi^2 + y^2} d\xi, \tag{4.12}$$

Following [1], one can show that system (2.13) is normally solvable provided that

$$\left(\frac{b}{a} - \frac{a}{b}\right)\left(\frac{\beta}{\alpha} - \frac{\alpha}{\beta}\right), \tag{4.13}$$

where $a \neq 0$, $b \neq 0$, $\alpha \neq 0$, $\beta \neq 0$.

Thus, the functions $\varphi(y, \lambda)$, $\psi(x, \lambda)$ and , consequently, the function v, given by formula (2.5), can be defined in a unique way. So, we have the following

Lemma. *Let one of conditions* (3.1) *be fulfilled together with one of conditions* (3.2). *Then, if inequality* (4.13) *takes place, and* $a \neq 0$, $b \neq 0$, $\alpha \neq 0$, $\beta \neq 0$, *then there exists the solution to problem* (1.17)–(1.20) *of the form* (2.5), *where the vector-function* $\vec{\Phi}^T = (\varphi, \psi)$ *is the solution to equation* (2.13).

Theorem 1. *Let conditions of lemma be fulfilled. Then there exists the solution to problme* (1.1)–(1.5) *and it is given by the formula:*

$$u(t, x, y) = \exp\left\{ -\frac{1}{2}(p_0 t - p_1 x - p_2 y) \right\} L^{-1} v(\lambda, x, y), \tag{4.14}$$

where L^{-1} *is the inverse Laplace transform, and the function* $v(x, y, \lambda)$ *is given by formula* (2.5).

5. Some special cases of the considered problem

Condition (4.13) can not be verified if, for example, $\alpha = 0$ or (and) $a = 0$. By this reason these cases are considered specially.

Theorem 2. *Let one of conditions* (3.1) *and* (3.2) *be fulfilled. If one of the relations*

a) $a = \alpha = 0$
b) $a = 0$, $\alpha \neq 0$, $\alpha \neq \beta$, (5.1)
c) $a \neq 0$, $\alpha = 0$, $a \neq b$

takes place, then there exists solution to problem (1.1)–(1.5).

The proof is similar to the proof of Theorem 1, and we will omit some details.

Let $a = 0$, $\alpha \neq 0$. We will reduce problem (1.1)–(1.5) to problem (1.17)–(1.20) and will seek the solution to this problem in the form:

$$v = V_{\mathbf{A}}^x \varphi + V_{\mathbf{A}}^y \psi + V_{\mathbf{B}}^y \psi. \tag{5.2}$$

Then instead of system (2.13) we come to

$$(\text{diag}[1, 2] + A)\vec{\Phi} = \vec{g}, \tag{5.3}$$

where diag[1, 2] is a 2×2 diagonal matrix with numbers 1 and 2 on its diagonal, elements of the operator matrix A are defined by the formulae:

$$a_{11} = 0, \quad a_{12} = B^x_{\mathbf{A}}(V^y_{\mathbf{A}} + V^y_{\mathbf{B}}),$$

$$a_{21} = (B^x_{\mathbf{A}} + B^y_{\mathbf{B}})V^x_{\mathbf{A}}, \quad a_{22} = B^y_{\mathbf{A}}V^y_{\mathbf{B}} + B^y_{\mathbf{B}}V^y\mathbb{A}.$$

The operators a_{kn} $(k, n = 1, 2)$ are of the form (4.5), where

$$a^0_{11} = 0, \quad a^0_{12} = \frac{b}{\beta}S_1 + \frac{b}{\alpha}S_2,$$

$$a^0_{21} = \frac{\beta}{b}S_1 + \frac{\alpha}{b}S_2, \quad a^0_{22} = \left(\frac{\beta}{\alpha} - \frac{\alpha}{\beta}\right)S,$$

and the singular operators S, S_1 and S_2 are given by formulae (4.5).

As a symbol of the opereator matrix A^0 with the elements a^0_{kn}, $(k, n = 1, 2)$ we use the 2×2 matrix function $S(t)$. Its elements can be found in accordance with [2]

$$s_{11}(t) = 0,$$

$$s_{12}(t) = \frac{1}{\sqrt{2}\mathrm{ch}\pi t}\left[\frac{b}{\beta}\left(\mathrm{ch}\frac{\pi t}{2} - i\mathrm{sh}\frac{\pi t}{2}\right) + \frac{b}{\alpha}\left(\mathrm{ch}\frac{\pi t}{2} + i\mathrm{sh}\frac{\pi t}{2}\right)\right],$$

$$s_{21}(t) = \frac{1}{\sqrt{2}\mathrm{ch}\pi t}\left[\frac{\beta}{b}\left(\mathrm{ch}\frac{\pi t}{2} - i\mathrm{sh}\frac{\pi t}{2}\right) + \frac{\alpha}{b}\left(\mathrm{ch}\frac{\pi t}{2} + i\mathrm{sh}\frac{\pi t}{2}\right)\right],$$

$$s_{22}(t) = \left(\frac{\beta}{\alpha} - \frac{\alpha}{\beta}\right)i\mathrm{th}\pi t.$$

Calculations show that

$$\Delta = \det(\mathrm{diag}[1, 2] + S(t)) = 2 + s_{22}(t) - s_{12}s_{21}(t) =$$

$$= 2 - \frac{1}{\mathrm{ch}\pi t}\left(\frac{1}{\mathrm{ch}\pi t} + \frac{\alpha^2 + \beta^2}{2\alpha\beta}\right) + i\left(\frac{\beta}{\alpha} - \frac{\alpha}{\beta}\right)\mathrm{th}\pi t.$$

The imaginary part of this deterinantcan turn to zero (for arbitrary α and β) at $t = 0$. The real part of Δ is positive if

$$2 - x^2(t) - 2Px(t) > 0, \tag{5.4}$$

where

$$x(t) = \frac{1}{\mathrm{ch}\pi t}, \quad (0 < x(t) \leq 1), \quad P = \frac{1}{4}\left(\frac{\alpha}{\beta} + \frac{\beta}{\alpha}\right).$$

Inequality (5.4) can take place iff

$$-P + \sqrt{P^2 + 2} > 0,$$

that is equavalent to the inequality $P < 0.5$. Hence the condition

$$\frac{\alpha}{\beta} - \frac{\beta}{\alpha} < 2.$$

immediately follows. If $t = 0$, then $\mathrm{Re}\Delta = 1 - 2P \neq 0$ for $P \neq 1/2$. Consequently, $\frac{\alpha}{\beta} + \frac{\beta}{\alpha} \neq 2$, that is equivalent to the condition b) of Theorem 2.

The case $\alpha = 0$, $a \neq 0$ is investigated in a similar way.

Finally, let $a = \alpha = 0$. The solution v of initial (1.17)–(1.20) is sought in the form

$$v = V_{\mathbf{A}}^{x}\varphi + V_{\mathbf{A}}^{y}\psi.$$

The determinant of the corresponding matrix symbol is

$$\det(I + S(t)) = 1 - \frac{1 - i\mathrm{ch}\pi t}{2\mathrm{ch}^2\pi t}.$$

As the parameter t varies this determinant describes a curve on the complex plane which does not contain the coordinate origin. So, the corresponding system of integral equations

$$(I + A)\vec{\Phi} = \vec{g}$$

is normally solvable.

Let us consider the initial–boundary value problem for the wave equation

$$c^{-2}w_{tt} - w_{xx} - w_{yy} = 0 \tag{5.5}$$

with boundary conditions

$$[w_t - aw_x - bw_y]\big|_{x=0} = q_1(y,t), \quad (y,t) \in \mathbb{R}_+^2, \tag{5.6}$$

$$[w_t - \alpha w_y - \beta w_x]\big|_{y=0} = q_2(x,t), \quad (x,t) \in \mathbb{R}_+^2, \tag{5.7}$$

and homogeneous initial equation (1.10) and the condition on edge (1.11).

Corollary 1. *Let $\beta > c$ and $b > c$. Then there exists the solution to mixed problem* (5.5)–(5.7), (1.10), (1.11) *if*

a) $\alpha \neq \beta$ at $a = 0$, $\alpha \neq 0$;
b) $a \neq b$, at $\alpha = 0$, $a \neq 0$;
c) $a = \alpha = 0$.

References

1. Dobrushkin V.A., On solvability of mixed problem for the wave equation in a domain with a corner. SJDE, V. 3, N. 2, 1997, pp. 143–150.
2. Dobrushkin V.A., Boundary value problems of dynamic theory of elasticity for wedge-shaped domains. Minsk, Nauka i teknika (Наука и техника), 1988, 416 pp.
3. Poruchikov V.B., Methods of dynamic theory of elasticity. Moscow, Nauka (Наука), 1986, 328 pp.

Translated by E.V.Mishchenko

Chapter 3

Ecosystem stability. 1. Predator-prey systems

L. V. Nedorezov[1]

Abstract. At present paper we consider the various estimations of the stability, diversity and fragility of ecosystems and apply its to analysis of the correlation between these estimations for different mathematical models of predator-prey systems.

Introduction

The estimation of ecosystems stability and biodiversity is among of the most important problems not only in modern science but in every day life, in forecast of human activity etc. In many cases our impossibility to estimate these main ecosystem characteristics, our impossibility to forecast the respective changing of these parameters led to catastrophic results in various regions of planet. That's why a lot of scientists all over the world try to construct the respective scales for ecosystems stability and diversity and to determine in the boundaries of their scales the results of applications of different methods which correlates with the influence on the ecosystem function [1–38].

There exist a lot of different publications which have the relation to the problem of estimation of the level of stability, diversity and fragility of the ecosystem [6, 8, 10, 12, 14, 15, 22, 24, 25, 27, 29, 32, 33, 37, 38]. Part of these publications has as a basic idea the R. MacArthur's hypothesis (see, for example, [27, 33] about the direct correlation between the stability of the biological society and it's diversity. The authors try to use different modifications of Fisher's or Shenon's measures [25, 27, 33] for obtaining the estimations of stability of ecosystems. Note that the estimations of such a kind can be applied as to the mathematical models of the ecosystem's dynamics as to the natural ecosystems and can be calculated independently. But in both cases we have a "one-moment value" and these estimations change with the changing of population's densities. It means that these estimations of stability level can change their values under the normal conditions (in particular, when we have the ordinary population's

[1]Novosibirsk State University, Novosibirsk, 630090, Russia

fluctuations) and when we haven't any changing in ecosystem stability. Also it means that we can't compare the different ecosystems between themselves with the help of the levels of it's stabilities or consider the changing of the stability level under the changing of the external influence of any density-independent factor on ecosystem (for example, as various industrial poisons, rate of flow in flowing ecosystems, middle year temperature etc.).

In other part of scientific investigations one has the applications of pure mathematical objects and ideas (like as Ljapunov's stability, existence of compact in phase space etc.) to the biological situations. In a result we have a set of various measures which can be determined for the mathematical models (sometimes it is very difficult to realize) and can't be applied to the analysis of natural ecosystems. We have it because all these estimations which we can find in modern mathematics were appeared under the solution of different problems of mechanics and physics and in reality doesn't correlate with basic ideas of stability which we have in modern ecology.

Finally, it is possible to point out another ways in obtaining the estimations of ecosystem stability and diversity. Below we'll consider one of these ways and we'll not try to apply any existing measures to ecological systems and take as a base any biological idea about the correlation between these measures which needs in separate examination. We assume that it is possible to find these estimations under the consideration of ecosystem perturbations under the changing of influence of external factor on the population's parameters.

The main direction in a constructing of new measures of the stability and diversity of ecosystems must assume the realization of the following requirements. First, it must be sufficiently universal and be calculated or efficiently estimated as for model as for natural conditions. Second, it must satisfy to ordinary imagine about the stability and allow us to compare the different regimes of the population dynamics between themselves with respect to it's level of stability.

At present paper we'll consider the first partial case when we have a "pure death factor" (i.e. it's influence leads only to increasing of death rates in populations; the similar situations we have, for example, for flowing ecosystems) and we'll demonstrate the possibilities of our approach with the help of some well-known systems. Another cases (and, in particular, the case with middle-year temperature when we have influence on all populations parameters) we'll consider in our following publications. Of course, the main idea of obtaining of the estimations will be preserved.

1. Estimations

Let's consider the following model of the ecosystem dynamics:

$$\frac{dx_k}{dt} = x_k f_k(x_1, \ldots, x_N) - \pi_k D x_k, \quad k = 1, \ldots, N \tag{1}$$

where $x_k(t)$ is a population density at time t, f_k is intensity of population changing, N is a number of species in ecosystem, D is a value of any external factor and it's influence leads to the additional death of the individuals in populations,

$D \in [0, D^*]$, $D^* = \text{const} > 0$, π_k is a "sensibility" of individuals to this factor D. Such kind of description of ecosystem dynamics we have for various models (see, for example, [1–5, 7–21, 23, 24, 26, 28, 30–36, 38] etc.). We'll assume that the rates f_k satisfy to the following conditions:

1. For all k functions f_k have the partial derivatives of all necessary orders.
2. In R_+^N of the phase space R^N there exist the stable invariant compact Δ and if initial point $\vec{x}_0 \in \Delta$ then for $\forall t\ \vec{x}(t) \in \Delta$.
3. For all k and for all points from R_+^N $|f_k| < \pi_k D^*$.

From these conditions we have that when $D = D^*$ in the phase space of the system (1) the origin is a global stable knot. It means that in this case all populations eliminate under all possible finite initial values of their number of individuals.

Following the Kolmogorov's idea [17, 18] we'll assume that (1) is a rough system. For every fixed value D in a phase space of the system (1) we have a set $A(D)$ of stable attractors. Every stable attractor $S \in A(D)$ determines the respective number $m(S)$ of the stable co-existence species (i.e. with non-zero values of it's densities for all $t > 0$ if their initial values are greater then zero). Respectively, $N - m(S)$ is a number of eliminated populations.

Let's consider two following functions:

$$M_1(D) = \max_{S \in A(D)} m(S), \quad M_2(D) = \min_{S \in A(D)} m(S). \tag{2}$$

Function $M_1(D)$ is equal to the maximum value of stable co-existent populations in ecosystem and $M_2(D)$ is equal to the minimum one: $\forall D\ 0 \leq M_2(D) \leq M_1(D) \leq N$.

When values of D increases we have the process of ecosystem destruction. At the fixed values D_j we have the "jumps" of functions $M_i(D)$ (2) that may correspond as to decreasing of number of stable co-existing species as to increasing of this number. Following by the ecological ideas the ecosystem stability we can estimate with the help of the following measures:

1. by minimum value of these "jumps" that corresponds to the first changing in species number variation:

$$S_1 = \min_k D_k = D_1$$

2. by maximum value among the "jumps" that corresponds to the extreme possibilities of ecosystem:

$$S_2 = \max_k D_k = D_m.$$

3. by middle value among the "jumps":

$$S_3 = \frac{1}{m} \sum_{k=1}^{m} D_k,$$

Obviously the last "jump" S_2 corresponds to the situation when under $\forall D \geq S_2$ all populations eliminate asymptotically under all non-negative initial values.

Elasticity of ecosystem we can estimate with the help of the following measures:

$$\begin{aligned} V_1 &= S_2 - S_1 \\ V_2^2 &= \frac{1}{m}\sum_{k=1}^{m}(D_k - S_3)^2, \end{aligned}$$

These measures have the following properties:

1. if $D_k \equiv D_1$ we have non-elastic society and $V_1 = V_2 = 0$.

2. if we have the changing of jump values $D_k^* = D_k + c$, where $c \equiv$ const, the elasticity doesn't change the value. In this case we have the changing only the values of stability.

3. if the values of jumps change in c times we have the respective changing of elasticity levels.

The diversity of ecosystem we can estimate with the help of the following integral:

$$\tilde{n} = \frac{1}{2D^*}\int_0^D (M_1(u) + M_2(u))du$$

This estimation is a mean value of species in ecosystem when $M_1 \equiv M_2$. In other cases it may be as any estimation of this mean value. It is necessary to note that it's very difficult to obtain the real mean value of species in ecosystem as in natural conditions as in mathematical model. For model we must know the number of stable attractors at every D, the measure of domains of their attraction etc.

Below we'll consider the case when $M_1 \equiv M_2$ which realizes for various models of predator-prey systems [1–5, 9–14 etc.] and, in particular, for Lotka-Volterra model.

It is necessary to note that this approach to the estimation of basic ecosystem characteristics allows us to obtain a base for the solution of the well-known ecological problem which was called as "stability vs. complexity". Really, for considering situation we can introduce two additional functions $L_1(D)$ and $L_2(D)$. Function $L_1(D)$ is equal to the maximum value of relation between the number of bonds (between populations) over the number of species in ecosystem. Respectively, $L_2(D)$ is equal to the minimum one. If the origin is a stable knot we'll assume that $L_2(D)$ or both functions (when it is global stable equilibrium) is equal to zero. So, solution of the problem correlates with the analysis of behavior of curves $(L_k(D), S_j(D))$ on the respective plane.

2. Verhulst's model

Let's consider the Verhulst model of isolate population dynamics [34]:

$$\frac{dx}{dt} = x(\alpha_1 - \alpha_2 - \beta x) - \pi Dx, \tag{3}$$

where $x(t)$ is a population size at time t, α_1 is a birth rate, α_2 is a death rate and β is a coefficient of self-regulative mechanisms act, $\varepsilon = \alpha_1 - \alpha_2$ is a Malthusian parameter.

As we assume $\varepsilon < \pi D^*$. For model (3) we have the following relations: $M_1 \equiv M_2$, $S_1 = S_2 = \varepsilon/\pi$ and $\tilde{n} = \varepsilon/(D^*\pi)$. It means that self-regulative mechanisms hasn't any influence on the stability of ecosystem and it's diversity. Stability of ecosystem and it's diversity are increasing under the growth of birth rate and are decreasing under the growth of death rate or growth of population sensibility π. So, in this case we have the linear correlation between stability of ecosystem and it's diversity (i.e. we consider the realization of R.MacArthur hypothesis).

3. Lotka-Volterra model of predator-prey system

Let's consider the following mathematical model of predator-prey system dynamics:

$$\frac{dx}{dt} = x(\alpha_1 - \alpha_2 - \beta x - \gamma_1 z) - \pi_1 Dx, \quad \frac{dz}{dt} = z(\gamma_2 x - \varepsilon - \sigma z) - \pi_2 Dz, \tag{4}$$

where all parameters are non-negative. For model (4) we also have the relation $M_1 = M_2$. We'll assume that predators don't eliminate when factor D has a sufficiently small value, i.e.:

$$\frac{\alpha_1 - \alpha_2}{\beta}\gamma_2 - \varepsilon > 0.$$

Realization of this inequality means that without the influence of considering factor we have two-species ecosystem, $M_1(0) = 2$. For model (4) we have the following "jumps" of factor D values:

$$D_1 = \frac{\gamma_2(\alpha_1 - \alpha_2) - \beta\varepsilon}{\beta\pi_2 + \gamma_2\pi_1}, \quad D_2 = \frac{\alpha_1 - \alpha_2}{\pi_1}.$$

Obviously, $D_1 < D_2$. So, for model (4) we have the relations:

$$S_1 = D_1, \quad S_2 = D_2;$$

$$\tilde{n} = \frac{1}{D^*}(D_1 + D_2).$$

It means that increasing of population's sensibilities π_1 and π_2 lead to decreasing of stability levels S_i and ecosystem diversity $\tilde{n}$. Also it is possible to point out that the respective behaviour of stability and diversity we have in a case with growth of strength of self-regulative mechanisms β and decreasing of Malthusian prey coefficient and growth of predators natural death rate ε. It's interesting that our estimations don't depend on the values of parameters γ_1 and σ that means that the respective mechanisms (i.e. the decreasing of preys number under the predation and self-regulation among the predators) haven't any influence on ecosystem stability and diversity.

So, for Lotka-Volterra model of predator-prey interaction (4) we have the realization of R.MacArthur hypothesis. The same results and the same character of influence of population parameters on the basic ecosystem characteristics we have for some other models of predator-prey system dynamics (in particular, for [1–5, 7–14, 16–21, 23, 24, 26, 28, 31–33, 35]).

4. Non-parametric model of predator-prey system

Let's consider the following non-parametric model of predator-prey system [8, 16, 23, 24] which has a sufficiently rich set of dynamic regimes. In particular, it describes the situations which correspond to the forest insect population outbreaks [8, 23]:

$$\frac{dx}{dt} = x(F(x) - f(x,z)) - Dx, \quad \frac{dx}{dt} = z(-G(z) + g(x,z)) - Dz, \tag{5}$$

where $x(t)$ and $z(t)$ are the densities of preys and predators respectively at time t. Functions F and G describe the self-regulative mechanisms act and functions f and g describe the interaction between populations. We'll assume that in considering case both populations have the equal values of sensibility with respect to the factor D. About the functions in the right-hand sides of equations (5) we'll assume that the following conditions are satisfying:

$$\begin{aligned}
&F(0) > 0 > F(\infty), \quad \frac{dF}{dx} \le 0, \quad \exists! x_T : F(x_T) = 0; \; F(0) < \infty;\\
&G(0) > 0, \quad \frac{dG}{dz} > 0, \quad G(\infty) = \infty;\\
&\forall (x,z) \in R_+^2 \; f, g \ge 0; \quad f(x,0) = g(0,z) = 0;\\
&\frac{\partial f}{\partial x} < 0, \quad \frac{\partial f}{\partial z} > 0, \quad \frac{\partial g}{\partial x} > 0, \quad \frac{\partial g}{\partial z} < 0;\\
&\exists K_1, K_2 \equiv \text{const} < \infty : \forall (x,z) \in R_+^2 \; f \le K_1, g \le K_2;\\
&K_2 = g(\infty, 0).
\end{aligned} \tag{6}$$

From these relations (6) we can obtain that in R_+^2 there exist stable invariant compact $\delta = [0, x_T] \times [0, z_T]$, where z_T is a solution of equation $G(z) = K_2$. If we have the realization of inequality $x^* > x_T$ where x^* is a solution of equation $g(x,0) = G(0)$ then in ecosystem we have both stable co-existing populations. Under the realization of inverse relation predators eliminate under all initial values of their density.

Under the increasing of values of D the limits of compact δ decrease monotonously: $\delta(D) = [0, \tilde{x}] \times [0, z_T]$ where $\tilde{x}$ is a solution of equation $F(x) = D$. And when equation $F(x) = D$ hasn't positive solutions then $M_1(D) = 0$. In other words we shall have the following relation:

$$\forall D : D \geq F(0) = D_2, \quad M_1(D) \equiv 0.$$

The amount of x^* increase monotonously under the increasing of D and when $D = K_2$ $x^*(D) = \infty$. It means that there exist a unique value D_1 which is a solution of equation $\tilde{x}(D) = x^*(D)$. The intersection of this value leads to the realization of situation when predators eliminate under all possible initial values of densities and ecosystem contains only one species.

So, we obtained that ecosystem diversity can be estimated as

$$\tilde{n} = \frac{1}{F(0)}(D_1 + D_2) = 1 + \frac{D_1}{F(0)}.$$

From this reation we can note that

1. ecosystem diversity increase under the increasing of effectiveness of transformation of "unit of preys to the unit of predators" (the same result we have, of course, for Lotka-Volterra model).

2. as we have the inequality $D_1 < K_2$ then under $F(0) \to \infty$ we have the asymptotic decreasing of ecosystem diversity; of course, it is possible to have the increasing of diversity under the growth of $F(0)$ as D_1 also increase under $F(0) \uparrow$.

3. diversity decreases under the growth of the role of self-regulative mechanisms in population of preys and doesn't depend on the strength of self-regulative mechanisms in population of predators.

For considering model we have the following values of ecosystem stabilities:

$$S_1 = D_1, \quad S_2 = D_2 = F(0), \quad S_3 = \frac{D_1 + D_2}{2} = \frac{1}{2}F(0)\tilde{n}$$

These relations shows if $F(0)$ doesn't change we have the linear relation between estimations of diversity and stability S_k: under the growth of diversity we have the growth of stability (for S_1 and S_3) or stability doesn't change (for S_2).

On the other side if we have the changing of Malthusian parameter (for example, from one geographical place to another) $F(0)$ then we can consider the following various situations:

1. diversity of ecosystem and it's stabilities have the inverse relations and diversity decreases (i.e. $\dfrac{D_1}{F(0)}$ decreases under $F(0) \uparrow$) under the growth of all estimations of ecosystem stability, i.e. $S_k \uparrow$ under $F(0) \uparrow$;

2. there is not a unique correspondence between diversity and stability of ecosystem (when the relation $\frac{D_1}{F(0)}$ can as growth as decreases under the increasing of values of $F(0)$) and in this case we can have several values of any stability measure to one value of diversity.

So, even for simple considered case we can obtain for every introduced measures of ecosystem stability and measure of it's diversity the different situations when hypothesis of R.MacArthur doesn't realize.

Estimation of elasticity for our model can be determined by the following expression:

$$V_1 = D_2 - D_1 = F(0)(2 - \tilde{n})$$

It means that under $F(0) = \text{const}$ the increasing of diversity leads to the decreasing of it's elasticity. Under $F(0) \to \infty$ and $F(0) >> D_1$ elasticity V_1 is a linear function of $F(0)$ and $V_1 \approx F(0)$.

It is necessary to note that changing of model parameters values (in particular, the value of Malthusian parameter $F(0)$) corresponds to the situation when we consider the similar ecosystems (in our case, the predator-prey systems) in various geographical places. In various places ecosystems have different values of their diversity and stability and comparison of the changing of these characteristics from the place to place must be in a base for examination of R. MacArthur hypothesis. On another words, analysis of the dependencies of ecosystem characteristics with respect to one or another population parameter has not only a simple sense but is a unique way for the examination of R. MacArthur hypothesis.

5. Conclusion

In modern literature we often can find the attempts to connect the idea of "stability of ecosystem" with traditional concepts about "stability" which we have in mathematics. Very often we can consider the same approach when one consider the mathematical model of ecosystem dynamics. But it is necessary to remember that modern mathematical conceptions about stability were appeared under the influence of physics and mechanics and solving the respective problems. And these problems and solutions haven't any relation to the ecological objects and can't correspond to the main conceptions about stability which appeared to this moment in modern ecology.

Really, the level of stability we can estimate with the help of measure (or volume) of the domain of it's stability (or with the help of relation of this volume to the volume of whole phase space) when the coming out the domain boundaries leads to the ecosystem destruction. Also we can estimate with the help of measure of domain in parameter space which correspond to the normal function of the system. It is possible to find a lot of another characteristics of such a kind which have any relation to the "stability". But the using of these characteristics has the serious difficulties even in very primitive cases with simple mathematical models. And for real model ecosystems it becomes a really great problem.

Moreover, some of existing estimations have a very simple scale as "stability exist — stability loosed". Such kind of scale doesn't allow us to compare the different ecosystems between themselves with respect to the levels of their stability and to analyze the process of stability changing under the growth of any external influence.

The estimations of stability, diversity and fragility of ecosystems which were considered at present paper have the following important properties. First, they haven't any hypothesis in a base (like Shenon's or Fisher's measures). The second, all estimations have a unique base and their values can be estimate under the analysis of a set of critical points which appear under the analysis of ecosystem destruction process. These critical values can be efficiently determined as for mathematical models (and it was demonstrated with the help of predator-prey models) as for real ecosystems. Moreover, these estimations can be obtained in the situation when we haven't any mathematical model of considering ecosystem.

And the third, this "non-model" character of estimations make them sufficiently universal and allow to use them practically in all models of ecosystem dynamics. In particular, their using can be useful under the determine of "ecological estimations" for the real role of one or another methods using in a practice.

References

1. F. Albrecht, H. Gatzke, A. Haddad, N. Wax, The dynamics of two interacting populations. J. of Math. Anal. and Appl., v. 46, 1974, pp. 658–670.
2. V. V. Alexeev, Influence of satiation factor on the predator-prey system dynamics. Biophysics, v. 18, n. 15, 1973, pp. 922–926.
3. V. V. Alexeev, Biophysics of ecosystems. UFN, v. 120, n. 4, 1976, pp. 647–676.
4. R. Arditi, J. M. Abillon, J. Vieira da Silva, A predator-prey model with satiation and intraspecific competition. Ecological Modelling, v. 5, n. 3, 1978, pp. 173–191.
5. A. D. Bazykin, Model of species dynamics and the problem of coexistence nearest species. J. of General Biology, v. 30, n. 3, 1969, pp. 259–264.
6. M. Bigon, J. Charper, K. Tausend, Ecology. Individuals, Populations and Societies. Moscow, Mir, v. 2, 1989, 278 p. (in Russian).
7. M. G. Bulmer, The theory of prey-predator oscillations. Theor. Pop. Biol., v. 9, 1976, pp. 137–150.
8. A. S. Isaev, R. G. Khlebopros, L. V. Nedorezov, Ju. P. Kondakov, V. V. Kiselev, Dynamics of Forest Insect Populations. Novosibirsk, Nauka, 1984, 224 p. (in Russian).

9. H. I. Freedman, Graphical stability, enrichment and pest control by a natural enemy. Math. Biocsi. v. 31, n. 3/4, 1976, pp. 207–225.

10. H. I. Freedman , I. W. H. So, Global stability and persistence of simple food chains. Math. Biosc., v. 76, n. 1, 1985, pp. 69–86.

11. H. I. Freedman, P. Waltman, Mathematical analysis of some three species food-chain models.- Math. Biosc., v. 33, n. 3, 1977, pp. 257–276.

12. T. C. Gard, Top predator persistence in differential equation models of food chains. J. Math. Biol., v. 14, n. 3, 1982, pp. 285–299.

13. N. S. Goel, S. C. Maitra, E. W. Montroll, On the Volterra and other nonlinear models of interacting populations. Rev. Modern Phys., v. 43, 1971, pp. 231–276.

14. K. Gopalsamy, Persistence in periodic and almost periodic Lotka-Volterra systems. J. Math. Biol., v. 21, n. 2, 1984, pp. 145–148.

15. V. Hutson, G. T. Vickers, A criterion for permanent coexistence of species, with an application to a two-prey one-predator system. Math. Biosc., v. 63, n. 2, 1983, pp. 253–269.

16. A. S. Isaev, R. G. Khlebopros, L. V. Nedorezov, Mathematical Model of Escape - Effect in Interaction of Preys and Predators. Mathematical Analysis of Forest Ecosystem Components, Novosibirsk, Nauka, 1979, pp. 74–82 (in Russian).

17. A. N. Kolmogoroff, Sulla theoria di Volterra della lutta per l'esistenza// Giornale dell'Inst Italiano degli attuary, v. 7, 1936, pp. 74–80.

18. A. N. Kolmogorov, Qualitative Analysis of Mathematical Models of Population Dynamics. Problems of Cybernetics, Moscow, Nauka, n. 25, 1972, pp. 101–106.

19. V. A. Kostitzin, La Biologie Mathematique. Paris, A.Colin, 1937, 236 p.

20. A. I. Lotka, Undamped oscillations derived from the law mass action. J. Amer. Chem. Soc., v. 42, n. 8, 1920, pp. 1595–1599.

21. A. I. Lotka, Elements of physical biology. Baltimore: Williams a. Wilkins, 1925, 460 p.

22. R. Margalef, Perspective in Ecological Theory. Chicago, 1968, 102 p.

23. L. V. Nedorezov, Modelling of Insect Population Outbreaks. Novosibirsk, Nauka, 1986, 125 p. (in Russian).

24. L. V. Nedorezov, Monofactor theory of the dynamics of biological populations. Krasnoyarsk, 1989, 28 p. (in Russian).

25. E. P. Odum, Basic Ecology. Moscow, Mir, 1975, 740 p.

26. R. A. Poluektov, Ju. A. Pikh, I. A. Shvitov, Dynamic models of ecological systems. Leningrad, Gydrometeoizdat, 1980, 288 p. (in Russian).

27. Yu. G. Puzachenko, Biodiversity, stability and function. Problems of Biological Systems Stability. Moscow, Nauka, 1992, pp. 5–32. (in Russian).

28. A. Rescigno, J. Richardson, The struggle for life: 1. Two species. Bull. Math. Bioph., v. 29, n. 2, 1967, pp. 377–388.

29. R. Ricklefs, The Economy of Nature. Moscow, Mir, 1979, 422 p.

30. M. L. Rosenzweig, Why the prey curve has a hump. Amer. Natur., v. 103, 1969, pp. 81–87.

31. M. L. Rosenzweig, R. H. MacArthur, Graphical representation and stability conditions of predator-prey interactions. Amer. Natur., v. 97, 1963, pp. 209–223.

32. J. M. Smith, Models in ecology. Moscow, Mir, 1976, 184 p.

33. Ju. M. Sviregev, D. O. Logofet, Stability of biological communities. Moscow, Nauka, 1978, 352 p. (in Russian).

34. P. F. Verhulst, Notice sur la loi que la population suit dans son accroissement. Corresp. Math. et Phys., n. 10, 1838, pp. 113–121.

35. V. Volterra, Leçons sur la theorie mathematique de la lutte pour la vie. Paris, Gauthiers-Villars, 1931, 214 p.

36. P. Waltman, Competition models in population biology. Philadelphia, 1987, 77 p.

37. K. Watt, Ecology and Rule of Natural Resources. Moscow, Mir, 1971, 463 p.

38. M. Williamson, The Analysis of Biological Populations. London, 1972, 268 p.

Translated by the autho1

Chapter 4

On the L^p boundedness of oscillatory integral operators

Ömer Akin[1] and I.A.Ikromov[2]

Abstract. In this work oscillatory integral operators are considered. We represented the solution of one of these problems of E.M.Stein on the L^p-boundednessof these operators. Both of the upper and lower estimations for norm era proved by large parameter of oscillation.

Introduction

The theory of pseudo-differential operators and Fourier integral operators are an important part of Modern mathematical physics [1–4]. However, many problems in both analysis and integral geometry lead to Fourier integral operators whose Lagrangianes are not local graph. For example, problems of diffraction had to led to operators with canonical folding relation [5–8].

Phond D.H. and Stein E.M. [6] considered oscillatory integral operators

$$(Tf)(x) = \int e^{i\lambda S(x,y)}\chi(y)f(y)dy, \quad t \in C_0^\infty(R),$$

where χ is a fixed C^∞ cut-off function on R with compact support, and phase function is a homogeneous polynomial

$$S(x,y) = \sum_{1\le j\le n-1} \alpha_{j-1}x^{n-j}y^j .$$

Phond D.H. and Stein E.M. proved the following theorem [7]:

Theorem. The operator T extends to a bounded operator on $L^2(R)$ with operator norm

$$||T|| \le C|\lambda|^{1/n},$$

when

(a) n is even and $|\alpha_{(n/2)-1}| + \ldots + |\alpha_0| \neq 0$;

(b) n is odd and $|\alpha_{((n-1)/2)-1}| + \ldots + |\alpha_0| \neq 0$.

[1]Univercity of Ankara, Department of Mathematics, Ankara, Turkey
[2]Samarkand Division of Uzbek Academy of Sciences, Samarkand, Uzbekistan

Stein E.M. [8] proposed a few open problems connected with Oscillatory Integral Operators. We represented solution of one of these problems.

We shall need a few definitions and notations in order to state the main result. Let Z_+ be the set of all non-negative integer numbers and R_+ be the set of all non-negative real numbers, Φ be the germ of a real analitic function and $\Phi(x,y) = \sum s_{nm}x^n y^m$ be the Maclaurin's series of the function Φ, where $(n,m) \in Z_+^2$. The set $\mathrm{supp}(\Phi) = \{(n,m) \in Z_+^2 \backslash \{(0,0)\} : s_{nm} \neq 0\}$ is called a support of Maclaurin's series of Φ. The convex hull of the set $\cup\big((n,m) + R_+^2\big)$, where $(n,m) \in \sup p(\Phi)$ is called Newton's polyhedron of the function Φ and denoted by $N(\Phi)$. Polyhedron $N(\Phi)$ is constructed by coordinates u and v. The coordinate of intersection point of the line $u = v$ with the boundary $N(\Phi)$ is called a distance to Newton's polyhedron and denoted by d. The polynomial $\Phi \in R[x,y]\backslash\{0\}$ is called a quasi-homogeneous polynomial of degree m with respect to weight $r = (r_1, r_2)$, (where $r_1 > 0$ and $r_2 > 0$) if for any $t > 0$ and $(x,y) \in R^2$ the following relation

$$f(t^{r_1}x, t^{r_2}y) = t^m f(x,y)$$

holds.

Let $L^p(R)$ be the space of all summability functions with degree p and $L^\infty(R)$ be the space of all bounded functions on R. The natural norm of the space $L^p(R)$ is denoted by $\|\cdot\|_p$ (where $1 \leq p \leq \infty$). We introduce the following class of amplitude

$$\Delta = \left\{\alpha \in L^\infty(R^2) \cap C^1(R^2) : \left\|\frac{\partial\alpha(x,.)}{\partial x}\right\|_\infty, \left\|\frac{\partial\alpha(.,y)}{\partial y}\right\|_\infty \in L^1(R)\right\}.$$

Definition. *Let $T : C_0^\infty \longmapsto L_{loc}^\infty(R)$ be a linear operator. The operator T is called an operator of type (p,q) if there exists a positive real number C such that for any $f \in C_0^\infty(R)$, the following inequality*

$$\|Tf\|_q \leq C\|f\|_p$$

holds.

It is clear that if T is an operator of type (p,q), then it has a bounded extention from $L^p(R)$ to $L^q(R)$. The extention's norm is denoted by $\|T\|_{p,q}$.

We consider oscillatory integral operators of the form

$$Z(T_\lambda f) = \int a(x,y)e^{i\lambda(x,y)\varphi(x,y)}f(y)dy,$$

where $a \in \Delta$ and $f \in C_0^\infty(R)$. The norm of the operator T_λ is unaffected by the addition to phase of terms $\varphi(x)$ or $\psi(y)$ and the next without loss of generality we assume $\Phi(x,0) \equiv 0$ and $\Phi(0,y) \equiv 0$.

It holds the following theorem:

Main Theorem. *Let $\Phi(x,y)$ be a quasihomogeneous polynomial, $d \in [1,2)$ be the distance to Newton's polyhedron, $p \in [1,2]$ and $\alpha \in \Delta$. Then the operator*

T_λ has the type of (p,q) (where $1/p+1/q=1$) and for the norm of $||T_\lambda||_{p,q}$ it holds the following asymtotical estimation

$$||T_\lambda|| = O\big(|\lambda|^{-1/pd}\big) \quad (as\ |\lambda| \to +\infty). \tag{2}$$

Moreover, if $\alpha(0,0) \neq 0$, then for the norm $||T_\lambda||_{p,q}$ it hollds the following relation

$$||T_\lambda|| \asymp O\big(|\lambda|^{-1/pd}\big) \quad (as\ |\lambda| \to +\infty). \tag{3}$$

Remark. The relation (3) means that, there exixt positive numbers c and C such that for any $|\lambda| \geq 1$ the following inequalities hold:

$$c|\lambda|^{-1/qd} \leq ||T_\lambda|| \leq C|\lambda|^{-1/qd}.$$

Our theorem is a generalization of the main theorem of [6].

This study consisits of two sections. The section 1 is devoted to L^p boundedness of some Oscillatory Integral Operators with special amplitudes. The section 2 is devoted to prove of the main theorem. Firstly, we shall prove the main theorem in particular cases $p=q=2$ and $p=1$, $q=\infty$, then we can apply the convexity theorem [12].

1. On the L^2 boundedness of Oscillatory Integral Operators

In this section, it is proved the lemma on L^2 boundedness of extension of Oscillatory Integral Operators with polynomial phases. First of all, we shall mention a few results required.

Let $F(x)$ be a real valued n times continuously differentiable function and α be a function of bounded variation on R. We consider the following trigonometric integral

$$J = \int \alpha(x)\exp(iF(x))dx. \tag{1.1}$$

Integral (1.1) is conditionally converged (see [3]). It holds the following generalized Van den Comput's lemma:

Lemma 1.1. *If $F^{(n)}(x)$ has no more than k zeros, then there exists a constant C depending only on n and k, such that the integral* (1.1) *has the estimation: $|J| \leq C\dfrac{||\alpha||_V}{H}$, where $||\alpha||_V = |\alpha(-\infty)| + V_R[\alpha]$, $V_R[a]$ is a total variation of α on R,*

$$H = \inf_{x\in R}\sum_{j=1}^{n}\left|F^{(j)}(x)\right|^{1/j}.$$

Some variants of the lemma 1.1 were proved in [13–14].

Let T be an integral operator with kernel $K(x,y)$. The following variant of Shur's lemma is proved bu the aid of the Schwarz's inequality [7].

Lemma 1.2. *Let there exists a measurable function $M(x,y)$ and non-negative numbers ε, C such that there hold the following inequalities:*

$$|K(x,y)| \leq M(x,y),$$

$$\int M(x,y)|x|^{-\varepsilon}dx \leq C|y|^{-\varepsilon},$$

$$\int M(x,y)|y|^{-\varepsilon}dy \leq C|x|^{-\varepsilon}.$$

Then the integral operator T is bounded on $L^2(R)$ and it's norm smoller that C.

Let $f \in C_0^\infty(R)$. We consider Oscillatory Integral Operators

$$(Tf)(x) = \int \exp\{i(y^2x + yx^m)\}f(y)dy. \tag{1.2}$$

It holds the followin lemma.

Lemma 1.3. *If $m \geq 2$, then the Oscillatory Integral Operator* (1.2) *has a bounded extension on $L^2(R)$.*

Proof. We show that for any $f \in C_0^\infty(R) \cap B_1$ it holds the inequality $||Tf|| \leq C$, where B_1 is a unit ball on $L^2(R)$ and C is a number depending only on m, $||\cdot||$ is the norm on $L^2(R)$.

We apply Lebesgue's theorem on convergence and have:

$$||Tf||^2 = \iint_K (x_1,x_2)f(x_1)f^*(x_2)dx_1dx_2,$$

where

$$K(x_1,x_2) = \int \exp\{iy^2(x_1 - x_2) + y(x_1^m - x_2^m)\}dy,$$

the sign "*" means complex conjugate of functions, hence the lemma 1.1 that the kernel $K(x_1,x_2)$ has the estimation

$$|K(x_1,x_2)| \leq C|x_1 - x_2|^{-1/2}$$

for any $(x_1,x_2) \in R^2$.

Next, the kernel $K(x_1,x_2)$ is represented as the sum of three functions

$$K(x_1,x_2) = K_1(x_1,x_2) + K_2(x_1,x_2) + K_3(x_1,x_2), \tag{1.3}$$

where

$$K_1(x_1,x_2) = K(x_1,x_2)X_{(x_1-x_2)\leq 1}(x_1,x_2),$$
$$K_2(x_1,x_2) = K(x_1,x_2)X_{(x_1-x_2)>1}(x_1,x_2),$$
$$K_3(x_1,x_2) = K(x_1,x_2)X_{(x_2-x_1)>1}(x_1,x_2),$$

and X is an indicator function of the according set. The integral operator T_1 with the kernel $K_1(x_1,x_2)$ has a bounded extension on $L^2(R)$ due to te lemma

1.2. We denote by T_2 an ntegral operator with the kernel $K_2(x_1, x_2)$ and prove of boundedness of T_2. It is easy to see that the kernel $K_2(x_1, x_2)$ is defined by the following expression:

$$K_2(x_1,x_2) = (x_1 - x_2)^{-1/2} X_{(x_1-x_2)>1}(x_1 - x_2) \exp\left\{ i\left(\frac{\pi}{4} + \frac{x_1^m - x_2^m}{4(x_- x_2)}\right)\right\}.$$

The kernel K_{22} of the operator $T_2^* T_2$ is expressed by the following oscillatory integral:

$$K_{22}(x_1, x_2) = \int \alpha(z, x_1, x_2) \exp\{iF(z, x_1, x_2)\} dz, \qquad (1.4)$$

where

$$F(z, x_1, x_2) = \frac{(z^m - x_1^m)^2}{4(z - x_1)} - \frac{(z^m - x_2^m)^2}{4(z - x_2)},$$

$$\alpha(z, x_1, x_2) = [(z - x_1)(z - x_2)]^{-1/2} X_{(z-x_1)>1}(z) X_{(z-x_2)>1}(z).$$

It is convenient to set $x_1 > 0$, $\xi = x_2/x_1$ and $\xi \in [-1, 1]$. These cases $x_1 < 0$ and also $|x_1/x_2| \le 1$ are considered by analogy. The integral (1.4) by changing of the variables $z \longmapsto x_1 z_1$ lead to the following oscillatoyr integral:

$$K_{22}(x_1, x_2) = \int \alpha_1(z, x_1, x_2, \xi) \exp\{i(x_1^{2m-1} F(z, 1, \xi))\} dz,$$

where

$$\alpha_1(z, x_1, x_2, \xi) = [(z - 1)(z - \xi)]^{1/2} X_{(z-1)>x_1^{-1}}(z) X_{(z-\xi)>x_1^{-1}}(z).$$

Let $\xi = \xi_0$ be a fixed point of the interval $[-1, 1]$. First of all, we consider the case $\xi_0 = ne1$. It is denoted by h_0 the function from the class $C_0^\infty(R)$, that is supported in the neighbourhood of points $z = 1$, $z = \xi_0$ and $h_0 - 1$ is vanished in some neighbourhood of these points. We consider the oscillatory integrals $K_0(x_1, x_2)$ and $K_\infty(x_1, x_2)$, defined by the equalities of

$$K_0(x_1, x_2) = \int \alpha_1 z, x_1, x_2, \xi) \exp\{i(x_1^{2m-1} F(z, 1, \xi))\} h_0(z) dz,$$

and

$$K_\infty(x_1, x_2) = K_{22}(x_1, x_2) - K_0(x_1, x_2).$$

Applying the lemma 1.2 to the oscillatory integral K_∞, then we have the estimation

$$|K_\infty| \le C|x|^{-1} X_{|\xi|\le 1}(\xi),$$

where $|x|$ is Euclide's distance of vector x. The lemma 1.2 implies that the integral operator T_∞ with the kernel K_∞ can be extended as a bounded operator on $L^2(R)$.

Next, we shall consider an estimation of the kernel $K_0(x_1, x_2)$. It should be noted that, if $\xi_0 \ne 1$, then the point $z = 1$ is not a point of singularities of phase function. In the case when m is an odd number and $m \ge 3$ there exists unique point $\xi = \eta \in (-1, 0)$ that the phase function $F(z, 1, \xi)$ has a singularity of type A_2 (see [2]). If $\xi_0 \neq \eta$, then the oscillatory integral $K_0(x_1, x_2)$ has the estimation:

$$\left|X_{|\xi-\xi_0|<\delta}(\xi)K_0(x_1, x_2)\right| \leq C(1+|x|)^{-(m-1)} = C(1+|x|)^{1-m},$$

in the case $\xi_0 = \eta$, it holds the inequality:

$$\left|X_{|\xi-\xi_0|<\delta}(\xi)K_0(x_1, x_2)\right| \leq C(1+|x|)^{-(4m-5)/6},$$

where δ is some small positive number.

Let $\xi_0 = 1$. The phase function $F(z, 1, \xi)$ is represented by the form $F(z, 1, \xi) = (\xi - 1)\varphi(z, \xi)$, where the function $\varphi(z, 1)$ at the point $z = 1$ has a non-degenerate critical point. Consequently, there exists a neighbourhood U of the point $\xi = 1$ such that for any $\xi \in U$ the oscillatory integral $K_0(x_1, x_2)$ has the estimation:

$$|K_0(x_1, x_2)| \leq C|\xi - 1|^{-3/4}|x|^{-(2m-2)/4}.$$

The kernel $K_3(x_1, x_2)$ is estimated by analogy.From these estimations due to lemma 1.2 we lead to prove of the lemma 1.3.

Let $f \in C_0^\infty(R)$. We consider the following Oscillatory Integral Operators:

$$(T_\lambda f)(x) = \int \alpha(y)\exp\{i\lambda(y^2x + yx^m)\}f(y)dy. \tag{1.5}$$

Lemma 1.4. *If $m \geq 2$ and $\alpha \in L^\infty(R)$, then the Oscillatory Integral Operators (1.5) has a bounded extension on $L^2(R)$. Moreover it's norm has the estimation*

$$||T_\lambda|| \leq C||\alpha||_\infty|\lambda|^{-1/d},$$

where $||\alpha||_\infty$ is the natural norm of the space $L^\infty(R)$ and C is some constant depending only on m.

Proof. We use the lemma 1.3 and have the following estimation:

$$||T_\lambda f||^2 = |\lambda|^{-(m-3)/(2m-1)}\int\left|\int \exp\{i(x^2y + xy^m)\}\alpha\left(\frac{y}{|\lambda|^{-1/(2m-1)}}\right)\times\right.$$

$$\left.\times f(y|\lambda|^{-1/(2m-1)})dy\right|^2 dx \leq$$

$$\leq C|\lambda|^{-(m+1)/(2m-1)}\int\left|\alpha(y|\lambda|^{-1/(2m-1)})f(y|\lambda|^{-1/(2m-1)})\right|^2 dy \leq$$

$$\leq C||\alpha||_\infty|\lambda|^{-(m+1)/(2m-1)}\int\left|f(y|\lambda|^{-1/(2m-1)})\right|^2 dy =$$

$$= C||\alpha||_\infty|\lambda|^{-m/(2m-1)}||f||^2.$$

The required estimation follows from here.

2. Prof of the main theorem

Firstly, we show that the operator T_λ has a type (2.2). Let the phase function has the form $\Phi(x,y) = px^2y + sxy^m$, where p, s are non-vanishing real numbers. Then the integral operator T_λ lead to Oscillatory Integral Operator with phase $\Phi = x^2y + xy^m$ by the aid of unitary transformations. According to the lemma 1.4, for any fixed real number $v \in R$ the integral operator

$$T_{\lambda v}(x) = \int_{-\infty}^{v} f(y)\exp\{i\lambda(x^2y + xy^m)\}dy$$

has the estimation

$$||T_{\lambda v}|| \leq C|\lambda|^{-1/d}, \tag{2.1}$$

where C is some constant depending only on m.

The relation (0.1) implies the following equality by applying integration by parts

$$||T_\lambda f||^2 = \int dx \left|\alpha(x,\infty)\int f(y)\exp\{i\lambda(x^2y + xy^m)\}dy\right|^2 +$$

$$+ \iint_d u\ dv \int dx \frac{\partial\alpha(x,u)}{\partial y}\int_{-\infty}^{u} f(y)\exp\{i\lambda(x^2\xi + x\xi^m)\}d\xi \times$$

$$\times \frac{\partial\alpha(x,v)}{\partial y}\int_{-\infty}^{u} f^*(\eta)\exp\{-i\lambda(x^2\eta + x\eta^m)\}d\eta.$$

From here by using the inequality (2.1) and Schwarz's inequality we have desired upper estimation of $||T_\lambda||_{2,2}$.

It holds the following lemma.

Lemma 2.1. *If $\alpha \in L^\infty(R^2)$, then the integral operator T_λ has a type $(1,\infty)$ and also the norm $||T_\lambda||_{1,\infty}$ uniformly bounded with respect to λ.*

Proof of the lemma 2.1 is evidently.

We can apply the convexity theorem of Riez-Tourin [12] and have:

$$||T_\lambda||_{p,q} \leq C|\lambda|^{-1qr}$$

for any $p \in [1,2]$.

Next we shall prove a lower estimation of the norm.

Lemma 2.2. *Let the phase function Φ of Oscillatory Integral Operator T_λ is a quasihomogeneous polynomial function of degree 1 with respect to weight (r_1, r_2) and the amplitude function $\alpha(x,y)$ is continuously at the point $(0,0)$, moreover $\alpha(0,0) \neq 0$. then for norm of Oscillatory Integral Operator T_λ holds the following lower estimation:*

$$||T_\lambda||_{p,q} \geq c|\lambda|^{-1/(r_1+r_2)q},$$

where c is some positive number.

Proof. It is convenient to set $\alpha(0,0)=1$. There exist positive numbers δ and γ such that for any $|x|+|y|<\delta$ and $|\lambda|\geq 1$ there hold the inequalities:

$$|\alpha(x,y)-1|<\frac{1}{4},$$

$$|\gamma\lambda^{-1}|^{r_1}+|\gamma\lambda^{-1}|^{r_1}\leq\delta,$$

$$\left|\int\limits_{-1}^{1}\int\limits_{-1}^{1}\exp\{i\gamma\Phi(x,y)\}dx\,dy\right|\geq\frac{1}{2}.$$

We consider the following family of functions depending on the parameter λ:

$$f_\lambda(x)=\left(\frac{|\lambda/\gamma|^{r_1}}{2}\right)^1/p\cdot X_{|x|\leq|\gamma/\lambda|^{r_1}}(x),$$

$$g_\lambda(x)=\left(\frac{|\lambda/\gamma|^{r_2}}{2}\right)^1/p\cdot X_{|x|\leq|\gamma/\lambda|^{r_2}}(x).$$

It is obvious that for any fixed $\lambda\neq 0$ the L^p norm of functions f_λ and g_λ is equal to 1.

We have the foolowing lower estimation for the norm of T_λ:

$$||T_\lambda||_{p,q}\geq|T_\lambda f_\lambda g_\lambda|=$$

$$=2^{-1/p}\left|\frac{\gamma}{\lambda}\right|^{(r_1+r_2)/q}\left|\int\limits_{-1}^{1}\alpha\left(x\left|\frac{\gamma}{\lambda}\right|^{r_1},y\left|\frac{\gamma}{\lambda}\right|^{r_2}\right)\exp\{i\gamma\Phi(x,y)\}dx\,dy\right|\geq$$

$$\geq 2^{-2-2/p}\left|\frac{\gamma}{\lambda}\right|^{(r_1+r_2)/q},$$

which is required to prove.

It should be noted that if the main edge of $N(\Phi)$ is compact then $D=(r_1+r_2)^{-1}$. In the case $d\in[1.2)$ the main edge of $N(\Phi)$ is compact. The proof of the lemma 2.2 is completed.

The upper estimation and the lemma 2.2 imply the proof of the main theorem.

References

1. Hormander L. Fourier integral operators, Acta Math. V. 127, 1971, pp. 79–183.
2. Airy G.B. On the intensity of light in neighbourhood of a caustic, Trans. Cambridge Philos. Soc., V. 6, 1938, pp. 379–403.
3. Melrose R., Taylor M. Near peak scattering and the correct Kirchhoff approximation for a convex obstacle, Adv. Math., V. 55, 1985, pp. 242–315.

4. Taylor M. Propagation, reflection and diffraction of singularities of solutions to wave equations, Bull. AMS, V. 84, 1978, pp. 589–611.

5. Palamodov V.P. Preprint, Matematisk ins., Universitet i Oslo, 1993.

6. Phong D.H., Stein E.M. Radon transforms and torision, Int. Math. Res. Notices, V. 4, 1991, pp. 49–60.

7. Phong D.H., Stein E.M. Oscillatory integrals with polynomial phases, Invent. Math., V. 110, 1992, pp. 39–62.

8. Stein E.M. Oscillatory integrals in Fourier analysis, Ann. of Math. Stud., Princeton Univ. press., N. 112, 1986, pp. 307–355.

9. Stein E.M. Problems in harmonic analysis related to curvature and oscillatory integrals, Int. Sim. Math. (ICM86), 1986, pp. 196–221.

10. Varchenko A.N. Newton's polyhedron and estimates of oscillatory integrals, Funct. anal. i ego. pril.(Функциональный анализ и его приложения), V. 10, N. 5, 1976, pp. 13–38 (in Russian).

11. Karpuchkin V.N. Theorem of the uniformly estimation of oscillatory integrals with phase depending on two variables, In book Tr. sem. im. I.G.Petrovskogo (Труды семинаров им. И.Г.Петровского), Moscow, Moscow State University, V. 10, 1984, pp. 150–169 (in Russian).

12. extension of a convexity theorem due to M.Riesz, Kungl. Fysiografiska Saellskapet i Lund Forhaendlinger 8, N. 14, 1939.

13. Van der Corput, Composition Math., V. 1, 1934, pp. 15–38.

14. Arkhipov G.I., Karasuba A.A., Chubarikov V.N., Theory of multiple trigonometrical sums, Moscow, Nauka, 1987 (in Russian).

Translated by the author

Chapter 5

The stability of numerical boundary treatments for finite-difference schemes for hyperbolic systems

A. M. Blokhin[1] and R. D. Alaev[2]

Abstract. Different aspects of the problem on influence of boundary conditions on stability of difference schemes used for calculation, for example, of gas dynamics problems are considred in the present paper.

Introduction

The present moment in science is characterized by wide distribution of the so-called methods of mathematical simulation. By the mathematical simulation (for example, in physics) we mean the following chain of actions while investigating a phenomenon: **describtion of the physical model of the phenomenon — creation of the mathematical model of the phenomenon — construction of the calcualtion model of phenomenon — creation of the calculating program — calculation with computer**.

A long-standing experience of solving practical problems from continuum mechanics enables us to conclude that conscious application of methods of mathematical simulation is not possible without a detailed analysis of relations between the mathematical and calculation models of the phenomenon. It should be noted that the approach based on simulteneous investigation of the initial mathematical problem and its finite-differential analog has been used in theory of differential equations for a long time. It is pertinent to note here , for example, the fundamental work [1] and also monographs [2–4] in which this approach was applied and developed.

As a basis in construction and investigation of calculation models we consider the requirement **of adequacy of the calculation model to the initial differential problem**. By adequacy we mean the following: the calculation model is constructed such that a theorem on existence of solution to the initial problem can be proved with its help. The latter seems to be an important fact since while calculating we must be sure that the approximate solutions tends in the limit to solution of the initial differential problem.

[1]Institute of Mathematics, Novosibirsk, 630090, Russia

[2]Bukhara State University, Bukhara, 705018, Uzbekistan

Some **initial mathematical models** were considered in the paper. First, **mixed problem for symmetric t-hyperbolic system with dissipative boundary conditions** was taken (see [2]).

Problem I. Find solution $\mathbb{U}$ to symmetric t-hyperbbolic (by Friedrichs) system:

$$A \cdot \mathbb{U}_t + B \cdot \mathbb{U}_x + C \cdot \mathbb{U}_y = 0 \quad \text{in} \quad R^3_{++}, \tag{0.1}$$

which satisfy boundary conditions at $x = 0$:

$$\mathbb{U}^{\mathbf{I}} = S \cdot \mathbb{U}^{\mathbf{II}} \tag{0.2}$$

and initial data at $t = 0$:

$$\mathbb{U}(0, x, y) = \mathbb{U}_0(x, y). \tag{0.3}$$

Here $A =$ block diag (A^I, A^{II}, A^{III}), $B =$ block diag $(I_{N_0}, -I_{N_1}, O_{N_2})$ are block diagonal matrices;

$$A^I = \text{diag}(a_1, a_2, \ldots, a_{N_0}),$$

$$A^{II} = \text{diag}(a_{N_0+1}, \ldots, a_{N_0+N_1}),$$

$A^{III} = \text{diag}(a_{N_0+N_1+1}, \ldots, a_N)$ are diagonal matrices; $a_i > 0$, $i = \overline{1, N}$; $N_0 + N_1 + N_2 = N$;
I_{N_0}, I_{N_1} are the unit matrices of order N_0, N_1 , correspondingly; O_{N_2} is zero square matrix of order N_2;

$$C = \begin{pmatrix} C_1 & C_2 & C_3 \\ C_2^* & C_4 & C_5 \\ C_3^* & C_4^* & C_6 \end{pmatrix},$$

and $C_1 = C_1^*$, $C_4 = C_4^*$, $C_6 = C_6^*$ are square matrix of order N_0, N_1, N_2 , correspondingly; C_2, C_3, C_5 are matrices of order $N_0 \times N_1$, $N_0 \times N_2$, $N_1 \times N_2$, correspondingly;

$$\mathbb{U} = \mathbb{U}(t, x, y) = \begin{pmatrix} u_1 \\ \vdots \\ u_N \end{pmatrix}, \quad \mathbb{U}^I = \begin{pmatrix} u_1 \\ \vdots \\ u_{N_0} \end{pmatrix},$$

$$\mathbb{U}^{II} = \begin{pmatrix} u_{N_0+1} \\ \vdots \\ u_{N_0+N_1} \end{pmatrix}, \quad \mathbb{U}^{III} = \begin{pmatrix} u_{N_0+N_1+1} \\ \vdots \\ u_N \end{pmatrix},$$

S is real constant matrix of order $N_0 \times N_1$,

$$R^3_{++} = \{(t, x, y);\ t, x > 0,\ y \in R^1\}.$$

Remark 0.1. Symmetric system (0.1) with real constant coefficients is given in the so-called **canonical form** (for details see [2]).

Let boundary conditions (0.2) be **strictly dissipative** (see [2]), i.e.

$$-(B\cdot\mathrm{U},\ \mathrm{U})|_{x=0} \ = \ (\mathrm{U}^{II},\ [I_{N_1}-S^*\cdot S]\cdot\mathrm{U}^{II})|_{x=0} \ \geq \ k_0\cdot(\mathrm{U}^{II},\mathrm{U}^{II})|_{x=0}, \quad (0.4)$$

where $k_0 > 0$ is a constant. Then, by [2], an identity of energy integral on smooth solution of system (0.1) can be written down (under fulfilment of (0.4) the identity is called **dissipative energy integral**):

$$\frac{\partial}{\partial t}(A\cdot\mathrm{U},\ \mathrm{U}) + \frac{\partial}{\partial x}(B\cdot\mathrm{U},\ \mathrm{U}) + \frac{\partial}{\partial y}(C\cdot\mathrm{U},\ \mathrm{U}) \ = \ 0. \qquad (0.5)$$

With its help for the case of dissipative boundary conditions (0.2) for **problem I** the following **a priori estimation** is valid:

$$J(t) \ \leq \ J(0), \quad t \ > \ 0, \qquad (0.6)$$

where

$$J(t) \ = \ \int\!\!\int_{R^2_+}(A\cdot\mathrm{U},\ \mathrm{U})dx\ dy, \ \ R^2_+ \ = \ \{(x,y);\ \ x>0,\ y\in R^1\}.$$

Coming back to reasoning on differential models for **problem I**althought this remark stays valid for other mathematical models, we will construct them such that they permit existence of **differential analog** of dissipative energy integral (0.5). Finally it bring the possibility to otbain energy estimation (differential analog of a priori estimate (0.6)) and thus to prove stability of the considered scheme. That will show adequacy of the differential model of initial **problem I** since, known the energy estimation, the theorem on existence of sufficiently smooth solution is proved by standard arguments (which the reader can find in [2]).

Remark 0.2. If the matrix C is diagonal the constant k_0 from (0.4) can be considered close to the unit.

As the second model we consider the mixed problem for the system of acoustics equations with boundary conditions on the shock wave taken from [4].

Problem II. Find solution U to system of acoustics equations

$$A\cdot\mathrm{U}_t + B\cdot\mathrm{U}_x + (C+\omega\cdot A)\cdot\mathrm{U}_y = 0 \ \ \text{в}\ R^3_{++}, \qquad (0.7)$$

which satisfies boundary conditions at $x = 0$:

$$u + d\cdot p = 0,\ \ v_t + \omega\cdot v_y - \lambda\cdot p_y = 0, \qquad (0.8)$$

and the initial data at $t = 0$:

$$\mathrm{U}(0,x,y) = \mathrm{U}_0(x,y). \qquad (0.9)$$

Here

$$A = \begin{pmatrix} 1 & 0 & 0 \\ 0 & M^2 & 0 \\ 0 & 0 & M^2 \end{pmatrix}, \quad B = \begin{pmatrix} 1 & 1 & 0 \\ 1 & M^2 & 0 \\ 0 & 0 & M^2 \end{pmatrix},$$

$$C = \begin{pmatrix} 0 & 0 & 1 \\ 0 & 0 & 0 \\ 1 & 0 & 0 \end{pmatrix}, \quad \mathrm{U} = \begin{pmatrix} p \\ u \\ v \end{pmatrix};$$

M $(0 < M < 1)$, $\omega \geq 0$, d, λ are constants. System (0.7) is obtained by linearization of quasilinear gas dynamics equations (see below) with respect to constant fundamental solution behind the oblique jump of sealing with special scaling of the linearized system, boundary conditions (0.8) are linearization of the Rankine-Hugoniot on the shock wave; u, v, p are small perturbations of the vector of velocity, pressure (see [4]). In some cases together with **problem II** we will consider the relations:

$$S_t + S_x + \omega \cdot S_y = 0 \quad \text{в } R^3_{++},$$

$$S + p = 0 \quad \text{at} \quad x = 0,$$

$$S(0, x, y) = S_0(x, y) \quad \text{at} \quad t = 0.$$

Here S are small perturbations of the entropy (see [4]).

Problem II is studied in details in [4]. The basic moment in investigation is the fact that **prolem II** can e reduced to the following auxiliary problem.

Problem III. We seek solution to the wave equaiton

$$\{M^2 \cdot (\hat{\tau} + \xi)^2 - \xi^2 - \eta^2\} p = 0 \text{ in domain } R^3_{++}, \tag{0.10}$$

which satisfy boundary condition at $x = 0$

$$\{M^2 \cdot (1 + d) \cdot \hat{\tau}^2 - \beta^2 \cdot \hat{\tau}\xi + M^2 \cdot \lambda\eta^2\} p = 0 \tag{0.11}$$

and initial data at $t = 0$

$$p(0, x, y) = p_0(x, y), \; \frac{\partial p}{\partial t}(0, x, y) = -\omega \cdot \frac{\partial p_0}{\partial y}(x, y) -$$

$$- \frac{\partial p_0}{\partial x}(x, y) - \frac{\partial u_0}{\partial x}(x, y) - \frac{\partial v_0}{\partial y}(x, y). \tag{0.12}$$

Here $\tau = \dfrac{\partial}{\partial t}$, $\xi = \dfrac{\partial}{\partial x}$, $\eta = \dfrac{\partial}{\partial y}$, $\hat{\tau} = \tau + \omega \cdot \eta$, $\beta^2 = 1 - M^2$.

Domains of ill-posedness of **problem III** are extracted by construction of **examples of ill-posedness of the Hadamard type** (see [2]). For this purpose we seek solutions to **problem III** in the form:

$$p = p^0 \cdot \exp\{i \cdot (-\hat{\omega} \cdot t + \hat{k} \cdot y + \hat{l} \cdot x)\}, \tag{0.13}$$

where p^0, $\hat{\omega}$, $\hat{k}$, $\hat{l}$ are constants, $i = \sqrt{-1}$. The constants $\hat{\omega}$, $\hat{k}$, $\hat{l}$ are related by the algebraic equations:

$$\begin{gathered} M^2(\hat{l} - \hat{\omega} + \omega \cdot \hat{k})^2 = \hat{l}^2 + \hat{k}^2, \\ M^2(1 + d)(-\hat{\omega} + \omega \cdot \hat{k})^2 - \beta^2(-\hat{\omega} + \omega \cdot \hat{k})\hat{l} + M^2 \cdot \lambda \cdot \hat{k}^2 = 0. \end{gathered} \tag{0.14}$$

Detailed analysis of (0.14) leads to the following conclusions (see [4]):

a). In domain $\mathcal{K}$: $\lambda < 0,\ d \cdot \beta^2 + M^2 \cdot \lambda > 0$ **problem III** has solution of the type (0.13) with $\text{Im}\hat{\omega} \leq 0,\ \text{Im}\hat{l} \geq 0,\ \text{Im}\hat{k} = 0$.

It was shown in [4] that domain $\mathcal{K}$ is the domain of well-posedness of **problem III** (and **problem II** simulteneously). Moreover, domain $\mathcal{K}$ is the domain where the **uniform Lopatinskii condition** is fulfilled. The latter means the well-posedness of **problem II** and the problems close to the initial one.

b) At other points of the plane either the uniform Lopatinskii condition or just Lopatinskii condition is not fulfilled. They are those points of the plane d, λ for which **problem III** has solutions of the type (0.13) with $\text{Im}\ \hat{\omega} > 0$, $\text{Im}\ \hat{l} < 0, \text{Im}\hat{k} = 0$. For the last case ill-posedness of **problem III** (and initial **problem II** also) is established by construction of the following Hadamard example

$$p_{\hat{k}} = p^0 \cdot \exp\{\sqrt{|\hat{k}|} + i \cdot (-\hat{\omega} \cdot t + \hat{k} \cdot y + \hat{l} \cdot x)\},\quad |\hat{k}| = 1, 2, \cdots.$$

Non-validity of the uniform Lopatinskii condition means that there exists a problem close to initial **problem II** such that the Hadamard example can be constructed ([4]).

The well-posedness of mixed **problem II** in the domain $\mathcal{K}$ is proved with the help of technique of dissipative energy of integrals allthought in a complicated case as compared with **problem I**. The matter is thar boundary conditions (0.8) are not strictly dissipative, i.e., for **problem II**

$$-(BU, U)|_{x=0} > 0$$

is not true.

By this reason this simle idea for obtaining a priori estimation (0.6) for **problem I** does not work here. In order to obtain a priori estimation for **problem II** an extended system (i.e., the sytem for the vector U and derivatives of U) was constructed, boundary conditions for such system will be strictly dissipative.

The extended system with its boundary conditions for **problem II** was constructed in [4]. Therewith, if $(d, \lambda) \in \mathcal{K}$ then the following a priori estimation takes place:

$$||\mathbb{U}(t)||_{W_2^2(R_+^2)} \leq C_1(T) \cdot ||\mathbb{U}_0||_{W_2^2(R_+^2)}, 0 < t \leq T < \infty. \tag{0.15}$$

Here $C_1(T) > 0$ is a constant finally dependent on T; W_2^2 is the Sobolev's space.

We note that if the point (d, λ) belongs to the domain where the uniform Lopatinskii condition does not hold, the following inequality is a priori estimation for **problem II**:

$$||\mathbb{U}(t)||_{W_2^2(R_+^2)} \leq C_2(T) \cdot ||\mathbb{U}_0||_{W_2^3(R_+^2)}, 0 < t \leq T < \infty. \tag{0.16}$$

Here $C_2(T) > 0$ is a constant. Estimation of the form (0.16) is called the estimation with the loss of smoothness ((0.15) is the estimation without the loss of smoothness).

The important moment while constructing the extended system is reduction of **mixed problem** III to the mixed problem for a symmetric t-hyperbolic (by Friedrichs) system. The iddea of reduction is as follows. If p satisfies (0.10) then the vector

$$\mathbf{Y} = \begin{pmatrix} \tilde{L}_1 \\ L_2 \\ \tilde{L}_3 \end{pmatrix} p$$

satisfies the system (see [4]):

$$\{D \cdot \hat{\tau} - Q \cdot \xi - R \cdot \eta\}\mathbf{Y} = 0, \tag{0.17}$$

Here

$$L_1 = \frac{1}{\beta^2}\hat{\tau}, \; L_2 = \xi - \frac{M^2}{\beta^2}\hat{\tau}, \; L_3 = \eta,$$

$$\tilde{L}_1 = ML_1, \; \tilde{L}_3 = \frac{1}{\beta}L_3, \; D = \frac{1}{\beta^2}(E + M^2 Q).$$

$$E = M \begin{pmatrix} 1 & m_1 & 0 \\ m_1 & 1 & 0 \\ 0 & 0 & 1 \end{pmatrix}, \; Q = \begin{pmatrix} m_1 & 1 & 0 \\ 1 & m_1 & 0 \\ 0 & 0 & -m_1 \end{pmatrix},$$

$$R = \begin{pmatrix} 0 & 0 & 1 \\ 0 & 0 & m_1 \\ 1 & m_1 & 0 \end{pmatrix},$$

m_1 is a constant. We note that $D > 0$ if $|m_1| < 1$.

Compose an extended system from two systems of the form (0.17):

$$\left\{\begin{pmatrix} D & O_3 \\ O_3 & D \end{pmatrix}\hat{\tau} - \begin{pmatrix} Q & O_3 \\ O_3 & Q \end{pmatrix}\xi - \begin{pmatrix} R & O_3 \\ O_3 & R \end{pmatrix}\eta\right\}\mathbf{X} = 0, \tag{0.18}$$

where

$$\mathbf{X} = \{\tilde{L}_1 L_+, \, L_2 L_+, \, \tilde{L}_3 L_+, \, \tilde{L}_1 L_-, \, L_2 L_-, \, \tilde{L}_3 L_-\}^*,$$

$$L_\pm = (a_1 L_1 + a_2^\pm L_2)p,$$

$$a_1 = \frac{m}{M}, \; a_2^\pm = -\frac{n}{a_\pm}, \; \gamma = \frac{\beta}{M^2}, \; n = -\frac{\lambda}{\beta},$$

$$a_\pm = \frac{M(\gamma \pm \sqrt{\gamma^2 - 4mn})}{2m}, \; m = \beta d + \frac{\lambda M^2}{\beta}, \; \gamma^2 > 4mn.$$

We note that in the domain $\mathcal{K} : m, n > 0$.

Boundary condition (0.11) is presented in the form:

$$\{\tilde{L}_1 - a_\pm L_2\}L_\pm = 0, \; x = 0. \tag{0.19}$$

Let $m_1 = \dfrac{a_+}{1+a_+^2}$ for the first system of the form (0.17) and $m_1 = \dfrac{a_-}{1+a_-^2}$ for the second system of the form (0.17). Then boundary conditons (0.19) are strictly dissipative because

$$\left(\begin{bmatrix} Q & 0_3 \\ 0_3 & Q \end{bmatrix}\mathbf{X},\ \mathbf{X}\right)\Bigg|_{x=0} = a_+(a_1 L_1 L_2 p \,+\, a_2^+ L_2^2 p)^2 \,+$$

$$+\,\frac{a_+}{1\,+\,a_+^2}(a_1 L_1 \tilde{L}_3 p \,+\, a_2^+ L_2 \tilde{L}_3 p)^2 \,+\, a_-(a_1 L_1 L_2 p \,+\, a_2^- L_2^2 p)^2 \,+$$

$$+\,\frac{a_-}{1\,+\,a_-^2}(a_1 L_1 \tilde{L}_3 p \,+\, a_2^- L_2 \tilde{L}_3 p)^2 \,>0$$

In [4] one can find another descriptions of systems (0.17), (0.18) which enable omitting the restriction $\gamma^2 \,>\, 4mn$.

As it was mentioned above for **problem II** the domain $\mathcal{K}\,:\quad \lambda\,<\,0$, $d\cdot\beta^2 + M^2\cdot\lambda > 0$ is the domain of fulfilment of the uniform Lopatinskii condition. So the estimations obtained for problems with constant coefficients are valid for problems with variable coefficients. Then with the help of the certain technique the local theorem on existence and uniqueness of the classic solution for quasilinear equaitons of gas dynamics behind a curvilinear shock wave was proved under assumption that at every point of the curvilinear shock wave at the moment $t=0$ the following equalities took place:

$$\lambda < 0,\ d\cdot\beta^2 + M^2\cdot\lambda > 0.$$

We note that basic results obtained for the case of constant coeffifents do not take place for the case of variable coefficients (and, consequently, for the quasilinear case) if there exist estimations of the type (0.16) (with the loss of smoothness). All these facts are described in [4].

We will construct differential schemes for **problem II** such that they will admit existence of difference analogs of a priori estimations (0.15).

Up to this moment we considered only initial mathematicals models since adequacy of a difference scheme to the initial nonlinear differential problem is difficult to establish by many reasons. Nevertheless we will consider an example of initial quasilinear differential problem.

We consider a system of equations which describe three-dimensional gas flow under assumption that the gas is nonviscous non heat conducting and is in the state of local thermodynamical equilibrium, i.e., there exsists the state equation. In the Cartesian coordinate system $\mathbf{x} = (x_1, x_2, x_3)$ the gas dymanics equations can be presented in the divergent form:

$$\rho_t \,+\, \mathrm{div}(\rho\cdot\mathbf{v}) \,=\, 0,$$

$$(\rho\cdot v_i)_t \,+\, \sum_{k=1}^{3}(\Pi_{ik})_{x_k} \,=\, 0,\ \ i = 1,2,3, \tag{0.20}$$

$$\left\{\rho \cdot \left(E + \frac{|\mathbf{v}|^2}{2}\right)\right\}_t + \operatorname{div}\left\{\rho \cdot \mathbf{v}\left(E + \frac{|\mathbf{v}|^2}{2} + p \cdot V\right)\right\} = 0.$$

Here ρ is the gas density, $\mathbf{v} = (v_1, v_2, v_3)^*$ is the gas velocity, $|v|^2 = (v, v)$, $\Pi_{ik} = \rho \cdot v_i \cdot v_k + p \cdot \delta_{ik}$ is the tensor of impuls flux density, p is the pressure, E is the internal energy, $V = \dfrac{1}{\rho}$ is the specific volume. Besides, the following thermodynamical identity holds true:

$$T \cdot dS = dE + p \cdot dV, \tag{0.21}$$

here T is the gas temperature, S is the entropyя. It follows from (0.21) that in particular

$$p = -\left.\frac{\partial E}{\partial V}\right|_S = \rho^2 \cdot \left.\frac{\partial E}{\partial \rho}\right|_S, \quad T = \left.\frac{\partial E}{\partial S}\right|_\rho.$$

Thus, by adition of the state equation

$$E = E(\rho, S)$$

we close system (0.20) which turns now into the system to determine the vector of unknowns

$$\mathbb{U} = \begin{pmatrix} \rho \\ S \\ \mathbf{v} \end{pmatrix}.$$

Finally, to system (0.20) we add the **additional conservation law**

$$(\rho \cdot S)_t + \operatorname{div}(\rho \cdot S \cdot \mathbf{v}) = 0, \tag{0.22}$$

valid on smooth solutions of system (0.20). We also note that in account of (0.22) equations system (0.20) on the smooth solutions is equavalent to the nondivergent system

$$\begin{gathered} \frac{1}{\rho \cdot c^2} \cdot \frac{dp}{dt} + \operatorname{div}\mathbf{v} = 0, \\ \frac{dS}{dt} = 0, \\ \rho \cdot \frac{d\mathbf{v}}{dt} + \nabla p = 0, \end{gathered} \tag{0.23}$$

here $\dfrac{d}{dt} = \dfrac{\partial}{\partial t} + (\mathbf{v}, \nabla), c^2 = (\rho^2 \cdot E_\rho)_\rho > 0$ is the squared velocity of sound in gas.

The Cauchy problem (0.20) was considered in [5]; here the theorem on local existence of the classical solution to this problem is proved.

1. Investigation of stability of a certain explicit two-layers different scheme

It was mentioned in **introduction** that investigation of stability for difference schemes to **mixed problem I** will be carried out by the technique of finite-difference analogs of dissipative energy integrals. In this section we will make acquaintance with its basic elements.

In the domain R^3_{++} we construct a difference scheme with the steps $\Delta t = \Delta$, $\Delta x = h_x$, $\Delta y = h_y$. Introduce the following notation:

$$\mathrm{U}_{ij}^k = \mathrm{U}_i = \mathrm{U} = \mathrm{U}(k \cdot \Delta, i \cdot h_x, j \cdot h_y), \quad k, i, |j| = 0, 1, \cdots;$$

$$\varphi \mathrm{U}_{ij}^k = \mathrm{U}_{ij}^{k+1} = \hat{\mathrm{U}}_i = \hat{\mathrm{U}}, \ \psi^{\pm 1} \mathrm{U}_{ij}^k = \mathrm{U}_{i\pm 1, j}^k,$$

$$\theta^{\pm 1} \mathrm{U}_{ij}^k = \mathrm{U}_{i, j\pm 1}^k, \quad \tau = \varphi - 1, \quad \xi = \psi - 1, \quad \eta = \theta - 1,$$

$$\bar{\xi} = 1 - \psi^{-1}, \quad \bar{\eta} = 1 - \theta^{-1},$$

$$\xi_0 = \xi + \bar{\xi}; \quad \hat{\xi} = \xi \cdot \bar{\xi} = \xi - \bar{\xi};$$

$$\eta_0 = \eta + \bar{\eta}; \quad \hat{\eta} = \eta \cdot \bar{\eta} = \eta - \bar{\eta}; \quad r_x = \frac{\Delta}{h_x}, \quad r_y = \frac{\Delta}{h_y}.$$

Following [4,6,7], we formulate the difference model of **mixed problem I**:

$$\left\{ A \cdot \tau + \frac{1}{2} r_x \cdot B \cdot \xi_0 + \frac{1}{2} r_y \cdot C \cdot \eta_0 - A \cdot (a \cdot \hat{\xi} + b \cdot \hat{\eta}) \right\} \mathrm{U} = 0,$$

$$k, |j| = 0, 1, \cdots, \quad i = 1, 2, \cdots,$$

$$\mathrm{U}_0^I = S \mathrm{U}_0^{II}, \quad k, |j| = 0, 1, \cdots, \tag{1.2}$$

$$\mathrm{U}_{ij}^0 = \mathrm{U}_0(ih_x, jh_y), \quad i, |j| = 0, 1, \cdots. \tag{1.3}$$

Here $a, b > 0$ are constants. It is shown from (1.2) that in order to determine the components of the vector $\hat{\mathrm{U}}_0$ it is necessary to pose **additional boundary conditions**. In our case they take the form:

$$\hat{\mathrm{U}}_0^{II} = \mathrm{U}_1^{II}, \quad \hat{\mathrm{U}}_0^{III} = \mathrm{U}_1^{III}, \quad k, |j| = 0, 1, \cdots. \tag{1.4}$$

We will prove stability of difference model (1.1)–(1.4).

Theorem 1.1. *Let*

$$r_x \leq \left[-\frac{1}{4} - 2a + \sqrt{\frac{1}{16} + 2a} \right] \cdot \min_{i=\overline{1, N_0+N_1}} a_i,$$

$$r_y \leq (\sqrt{b} - 2 \cdot b) \cdot \frac{\nu_0}{\nu_C}, \tag{1.5}$$

$\nu_0 = \min_{i=\overline{1,N}} a_i$, $\nu_C = \|C\|$ *is an operator norm of the matrix* C,

$$a, b < \frac{1}{4}.$$

Then difference model (1.1)–(1.4) is stable in the energy norm $\sqrt{J_k}$, *here*

$$J_k = \tilde{J}_k + h_x \cdot h_y \cdot \sum_{j=-\infty}^{\infty} \{a \cdot (\mathbb{U}_0^{III}, A^{III} \cdot \mathbb{U}_0^{III}) + \frac{1}{2} \cdot (\mathbb{U}_0^{II}, B_+^{II} \cdot \mathbb{U}_0^{II})\},$$

$$B_+^{II} = 2 \cdot a \cdot A^{II} + r_x \cdot I_{N_1}.$$

$$\tilde{J}_k = h_x \cdot h_y \cdot \sum_{i=1}^{\infty} \sum_{j=-\infty}^{\infty} (\mathbb{U}, A\mathbb{U}).$$

Proof. Multiply (1.1) scalarly by $2\mathbb{U}$ and obtain a difference analog of the energy integral identity (see **introduction**):

$$\tau(\mathbb{U}, A \cdot \mathbb{U}) + \frac{r_x}{2} \cdot \xi_0(\mathbb{U}, B \cdot \mathbb{U}) + \frac{r_y}{2} \cdot \eta_0(\mathbb{U}, C \cdot \mathbb{U}) -$$

$$- a \cdot \hat{\xi}(\mathbb{U}, A \cdot \mathbb{U}) - b \cdot \hat{\eta}(\mathbb{U}, A \cdot \mathbb{U}) - \frac{r_x}{2} \cdot \xi(\bar{\xi}\mathbb{U}, B \cdot \bar{\xi}\mathbb{U}) -$$

$$- \frac{r_y}{2} \cdot \eta(\bar{\eta}\mathbb{U}, C \cdot \bar{\eta}\mathbb{U}) - (\tau\mathbb{U}, A \cdot \tau\mathbb{U}) + a \cdot (\bar{\xi}\mathbb{U}, A \cdot \bar{\xi}\mathbb{U}) +$$

$$+ a \cdot (\xi\mathbb{U}, A \cdot \xi\mathbb{U}) + b \cdot (\bar{\eta}\mathbb{U}, A \cdot \bar{\eta}\mathbb{U}) + b \cdot (\eta\mathbb{U}, A \cdot \eta\mathbb{U}) = 0. \qquad (1.6)$$

The following evident relations were used in deduction of (1.6):

$$\begin{cases} (2\mathbb{U}, B \cdot \xi\mathbb{U}) = \xi(\mathbb{U}, B \cdot \mathbb{U}) - (\xi\mathbb{U}, B \cdot \xi\mathbb{U}), \\ (2\mathbb{U}, B \cdot \bar{\xi}\mathbb{U}) = \bar{\xi}(\mathbb{U}, B \cdot \mathbb{U}) + (\bar{\xi}\mathbb{U}, B \cdot \bar{\xi}\mathbb{U}), \\ (2\mathbb{U}, A \cdot \hat{\xi}\mathbb{U}) = \hat{\xi}(\mathbb{U}, A \cdot \mathbb{U}) - (\bar{\xi}\mathbb{U}, A \cdot \bar{\xi}\mathbb{U}) - (\xi\mathbb{U}, A \cdot \xi\mathbb{U}), \end{cases}$$

an so on. Multiply (1.6) by $h_x h_y$ and sum up this expression with respect to i (from 1 to $+\infty$) and j (from $-\infty$ to $+\infty$):

$$\tilde{J}_{k+1} - \tilde{J}_k + h_x \cdot h_y \cdot \sum_{j=-\infty}^{\infty} \left\{ \frac{1}{2} r_x [-(\mathbb{U}_0, B \cdot \mathbb{U}_0) - (\mathbb{U}_1, B \cdot \mathbb{U}_1)] + \right.$$

$$\left. + a[-(\mathbb{U}_0, A \cdot \mathbb{U}_0) + (\mathbb{U}_1, A \cdot \mathbb{U}_1)] + \frac{1}{2} r_x (\xi\mathbb{U}_0, B \cdot \xi\mathbb{U}_0) \right\} +$$

$$+ h_x \cdot h_y \cdot \sum_{i=1}^{\infty} \sum_{j=-\infty}^{\infty} \{- (\tau\mathbb{U}, A \cdot \tau\mathbb{U}) +$$

$$+ a \cdot (\bar{\xi}\mathbb{U}, A \cdot \bar{\xi}\mathbb{U}) + a \cdot (\xi\mathbb{U}, A \cdot \xi\mathbb{U}) +$$

$$+ b \cdot (\bar{\eta}\mathbb{U},\ A \cdot \bar{\eta}\mathbb{U}) + b \cdot (\eta\mathbb{U},\ A \cdot \eta\mathbb{U})\} = 0. \tag{1.7}$$

Transform (1.7) in the following way: since

$$-(\tau\mathbb{U},\ A \cdot \tau\mathbb{U}) =$$

$$= -\left(A^{-1} \cdot \left[\left(\frac{1}{2}r_x\right) \cdot B \cdot \xi_0 + \left(\frac{1}{2}r_y\right) \cdot C \cdot \eta_0 - A \cdot (a \cdot \hat{\xi} + b \cdot \hat{\eta})\right]\mathbb{U},\right.$$

$$\left.\left[\left(\frac{1}{2}r_x\right) \cdot B \cdot \xi_0 + \left(\frac{1}{2}r_y\right) \cdot C \cdot \eta_0 - A \cdot (a \cdot \hat{\xi} + b \cdot \hat{\eta})\right]\mathbb{U}\right) \geq$$

$$\geq -\{(\xi\mathbb{U},\ B_- \cdot +(\bar{\xi}\mathbb{U},\ B_+ \cdot A^{-1} \cdot B_+ \cdot \bar{\xi}\mathbb{U}) +$$

$$+(\eta\mathbb{U},\ C_- \cdot A^{-1} \cdot C_- \cdot \eta\mathbb{U}) + (\bar{\eta}\mathbb{U},\ C_+ \cdot A^{-1} \cdot C_+ \cdot \bar{\eta}\mathbb{U})\},$$

(1.7) together with (1.2) gives

$$\tilde{J}_{k+1} - \tilde{J}_k + h_x \cdot h_y \cdot \sum_{j=-\infty}^{\infty} \left\{ \left(\frac{1}{2}r_x\right)(\mathbb{U}_0^{II},\ [I_{N_1} - S^* \cdot S]\mathbb{U}_0^{II}) - \right.$$

$$\left. -\left(\frac{1}{2}r_x\right)(\mathbb{U}_1, B \cdot \mathbb{U}_1) + a[-(\mathbb{U}_0,\ A \cdot \mathbb{U}_0) + (\mathbb{U}_1, A \cdot \mathbb{U}_1)] \right\} +$$

$$+h_x \cdot h_y \cdot \sum_{i=1}^{\infty} \sum_{j=-\infty}^{\infty} \{(\xi\mathbb{U},\ K_1 \cdot \xi\mathbb{U}) + (\bar{\xi}\mathbb{U}, K_2 \cdot \bar{\xi}\mathbb{U}) +$$

$$+(\eta\mathbb{U}, K_3 \cdot \eta\mathbb{U}) + (\bar{\eta}\mathbb{U}, K_4 \cdot \bar{\eta}\mathbb{U})\} \leq 0. \tag{1.8}$$

Here

$$K_1 = a \cdot A - B_- \cdot A^{-1} \cdot B_- - \frac{1}{2}r_x \cdot D_1,$$

$$K_2 = a \cdot A - B_+ \cdot A^{-1} \cdot B_+,$$

$$K_3 = a \cdot A - C_- \cdot A^{-1} \cdot C_-,$$

$$K_4 = a \cdot A - C_+ \cdot A^{-1} \cdot C_+,$$

$$B_\pm = 2a \cdot A \pm r_x \cdot B,$$

$$C_\pm = 2b \cdot A \pm r_y \cdot C,$$

$$D_1 = \text{block diag}\,(O_{N_0},\ I_{N_1},\ O_{N_2}).$$

If (1.5) are true, then

$$K_{1,2,3,4} \geq 0.$$

Then (1.8) transformes into

$$J_{k+1} - J_k + h_x \cdot h_y \cdot \sum_{j=-\infty}^{\infty} \left\{ \left(\mathbb{U}_0^{II}, \left[r_x \cdot I_{N_1} - \left(\frac{1}{2}r_x + a \cdot \nu\right) S^* \cdot S \right] \cdot \mathbb{U}_0^{II} \right) + \right.$$

$$+\frac{1}{2}(\mathrm{U}_1^I, B_-^I \cdot \mathrm{U}_1^I)\Big\} \le 0, \tag{1.8'}$$

where

$$B_-^I = 2a \cdot A^I - r_x \cdot I_{N_0}, \quad \nu = \max_{i=\overline{1,N}} a_i.$$

By (1.5), $B_-^I \ge 0$. Besides,

$$r_x \cdot I_{N_1} - \left(\frac{1}{2}r_x + a \cdot \nu\right) S^* \cdot S \ge \left[r_x - \left(\frac{1}{2}r_x + a \cdot \nu\right)(1 - k_0)\right] \cdot I_{N_1} \ge 0,$$

at

$$k_0 \ge \frac{a\nu - (r_x/2)}{a\nu + (r_y/2)}$$

(if C is a diagonal matrix, this equation is fulfilled a fortiori by **remark 0.2** from **introduction**). With account of this inequality, the desired a priori estimation (the difference analog of the a priori estimation (0.6) from **introduction**) follows from (1.8′):

$$J_{k+1} \le J_k, \text{ i.e. } J_k \le J_0, \quad k = 1, 2, \cdots. \tag{1.9}$$

It is known that namely inequality (1.9) points at stability of scheme (1.1) (in a view of the boundary conditions) in the energy norm $\sqrt{J_k}$.

Remark 1.1. The technique how to construct finite-difference analogs of dissipative energy integrals for a certain class of explicit two-layers difference schemes is given in details in survey [8] also. The case of variable coefficients is studied in [4,6].

2. Investigation of staility of a certain explicit three-layers different scheme

Let us consider an example of difference scheme for **mixed problem I**, its stability is established with the technique of finite-difference analogs of dissipative energy integrals.

We formulate the difference scheme for **problem I**:

$$A \cdot \tau_0 \mathrm{U} + r_x \cdot B \cdot \xi_0 \mathrm{U} + r_y \cdot C \cdot \eta_0 \mathrm{U} = 0,$$

$$k,\ i = 1,\ 2, \cdots,\ |j| = 0,\ 1, \cdots; \tag{2.1}$$

$$\mathrm{U}_0^I = 2 \cdot S \cdot \mathrm{U}_0^{II} - \frac{\hat{\mathrm{U}}_1^I + \check{\mathrm{U}}_1^I}{2},$$

$$k = 2,\ 3, \cdots, \quad |j| = 0,\ 1, \cdots; \tag{2.2}$$

$$\mathrm{U}_{ij}^0 = \mathrm{U}_0(i \cdot h_x,\ j \cdot h_y), i,\ |j| = 0,\ 1, \cdots. \tag{2.3}$$

Here $\tau_0 = \varphi - \varphi^{-1}$, $\mathrm{U}_{ij}^{k-1} = \check{\mathrm{U}} = \check{\mathrm{U}}_i$. It follows from (2.2) in order to determine the components of the vector U_0, $k = 2, 3, \cdots$, $|j| = 0, 1, \cdots$, it is necessary to pose additonal boundary conditions. We formulate them as follows:

$$\mathrm{U}_0^{II} = \frac{\hat{\mathrm{U}}_1^{II} + \check{\mathrm{U}}_1^{II}}{2}, \quad \mathrm{U}_0^{III} = \check{\mathrm{U}}_1^{III},$$

$$k = 2, 3, \cdots, \quad |j| = 0, 1, \cdots. \tag{2.4}$$

We note that in a view of difference scheme (2.1) boundary conditons (2.2) and additional boundary conditons (2.4) can be rewritten in the form:

$$\mathrm{U}_0^I = 2 \cdot D_I^{-1} \cdot S \cdot \mathrm{U}_0^{II} - D_I^{-1} \cdot \check{\mathrm{U}}_1^I + \frac{r_x}{2} \cdot D_I^{-1} \cdot R^I \cdot \mathrm{U}_2^I + \frac{r_y}{2} \cdot D_I^{-1} \cdot R^I \cdot K^I \cdot \eta_0 \mathrm{U}_1,$$

$$\mathrm{U}_0^{II} = D_{II}^{-1} \cdot \check{\mathrm{U}}_1^{II} + \frac{r_x}{2} \cdot D_{II}^{-1} \cdot R^{II} \cdot \mathrm{U}_2^{II} - \frac{r_y}{2} \cdot D_{II}^{-1} \cdot R^{II} \cdot K^{II} \cdot \eta_0 \mathrm{U}_1,$$

here

$$R^I = (A^I)^{-1}, \quad R^{II} = (A^{II})^{-1},$$

$$D_I = I_{N_0} + \frac{r_x}{2} \cdot R^I, \quad D_{II} = I_{N_1} + \frac{r_x}{2} \cdot R^{II},$$

$K^I = (C_1,\ C_2,\ C_3)$, $K^{II} = (C_2^*,\ C_4,\ C_5)$ are rectangular matrices of order $N_0 \times N$, $N_1 \times N$, correspondingly (matrices C_1, C_2, C_3, C_4, C_5 can be found in **introduction**).

Theorem 2.1. *Let*

$$r_x + r_y \cdot \nu_C < \nu_0. \tag{2.5}$$

be valid. Then difference scheme (2.1)–(2.4) *is stable in the energy norm* $\sqrt{J_k}$, *where*

$$J_k = h_x \cdot h_y \cdot \sum_{i=1}^{\infty} \sum_{j=-\infty}^{\infty} \{(\hat{\mathrm{U}},\ A \cdot \hat{\mathrm{U}}) + (\mathrm{U},\ A \cdot \mathrm{U}) +$$

$$+ r_y(\hat{\mathrm{U}},\ C \cdot \eta_0 \mathrm{U}) + r_x(\psi\hat{\mathrm{U}},\ B \cdot \xi_0 \psi \mathrm{U})\} + \Delta h_y \cdot \sum_{j=-\infty}^{\infty} (\hat{\mathrm{U}}_1,\ B \cdot \mathrm{U}_2).$$

Proof. Multiply system (2.1) scalarly by $(\hat{\mathrm{U}} + \check{\mathrm{U}})$:

$$\tau_0(\mathrm{U},\ A \cdot \mathrm{U}) + r_x(\hat{\mathrm{U}},\ B \cdot \xi_0 \mathrm{U}) + r_x(\check{\mathrm{U}},\ B \cdot \xi_0 \mathrm{U}) +$$

$$+ r_y(\hat{\mathrm{U}},\ C \cdot \eta_0 \mathrm{U}) + r_y(\check{\mathrm{U}},\ C \cdot \eta_0 \mathrm{U}) = 0. \tag{2.6}$$

The following easy-to-check relation was used while deducing (2.6):

$$(A \cdot \tau_0 \mathrm{U},\ [\hat{\mathrm{U}} + \check{\mathrm{U}}]) = \tau_0(A \cdot \mathrm{U},\ \mathrm{U}).$$

Multiply equality (2.6) by $h_x \cdot h_y$ and sum up this expression with respect to i (from 1 to $+\infty$) and j (from $-\infty$ to $+\infty$):

$$J_k - J_{k-1} - \Delta \cdot h_y \cdot \sum_{j=-\infty}^{\infty} \{(\mathrm{U}_0,\ B \cdot \check{\mathrm{U}}_1) + (\mathrm{U}_0,\ B \cdot \hat{\mathrm{U}}_1)\} = 0. \tag{2.7}$$

To infer (2.7) we used these simple relations:

$$\sum_{i=1}^{\infty}(\mathbb{U},\ B\cdot\xi_0\check{\mathbb{U}})\ +\ (\check{\mathbb{U}},\ B\cdot\xi_0\mathbb{U})\ =-\,(\mathbb{U}_1,\ B\cdot\check{\mathbb{U}}_0)\ -\ (\mathbb{U}_0,\ B\cdot\check{\mathbb{U}}_1),$$

$$\sum_{j=-\infty}^{\infty}\{(\mathbb{U},\ C\cdot\eta_0\check{\mathbb{U}})+(\check{\mathbb{U}},\ C\cdot\eta_0\mathbb{U})\}\ =0.$$

In a view of (2.4), strict dissipativeness of the boundary solutions (see inequality (0.4) from **introduction**) we come to

$$\begin{aligned}-(\omega,\ B\cdot\mathbb{U}_0)\ &=\ (\mathbb{U}_0^{II},\ \omega^{II})\ -\ (\mathbb{U}_0^{I},\ \omega^{I})\ =\\ &=\ (\mathbb{U}_0^{II},\ \omega^{II})\ -\ ([2S\cdot\mathbb{U}_0^{II}\ -\ \frac{1}{2}\omega^{I}],\ \omega^{I})\ =\\ &=\ \frac{1}{2}(\omega^{II},\ \omega^{II})\ +\ \frac{1}{2}(\omega^{I},\ \omega^{I})\ -\ (S\cdot\omega^{II},\ \omega^{I})\ \geq\\ &\geq\ \frac{1}{2}(\omega^{II},\ [I_{N_1}\ -\ S^*S]\cdot\omega^{II})\ \geq\ \frac{1}{2}k_0(\omega^{II},\ \omega^{II}).\end{aligned}\tag{2.8}$$

Here $\omega\ =\ \check{\mathbb{U}}_1\ +\ \hat{\mathbb{U}}_1,\ \omega^{I,II}\ =\ \check{\mathbb{U}}_1^{I,II}\ +\ \hat{\mathbb{U}}_1^{I,II}$. We also note that while deducing (2.8) we used the inequalities:

$$\begin{aligned}|(S\cdot\omega^{II},\ \omega^{I})|\ &\leq\ (S^*S\cdot\omega^{II},\ \omega^{II})^{1/2}\cdot(\omega^{I},\omega^{I})^{1/2}\ \leq\\ &\leq\ \frac{1}{2}(S^*S\cdot\omega^{II},\ \omega^{II})\ +\ \frac{1}{2}(\omega^{I},\omega^{I}).\end{aligned}$$

(2.8) and (2.7) result at

$$J_k\ \leq\ J_{k-1},\ \text{i.e.}\ J_k\ \leq\ J_0,\quad k\ =\ 1,2,\ldots\ .$$

As

$$|(\hat{\mathbb{U}},C\cdot\eta_0\mathbb{U})|\leq\nu_C\cdot(\hat{\mathbb{U}},\hat{\mathbb{U}})+\frac{1}{2}\nu_C\cdot\{(\theta\mathbb{U},\theta\mathbb{U})+(\theta^{-1}\mathbb{U},\theta^{-1}\mathbb{U})\},$$

$$|(\psi\hat{\mathbb{U}},B\cdot\xi_0\psi\mathbb{U})|\leq(\psi\hat{\mathbb{U}},\psi\hat{\mathbb{U}})+\frac{1}{2}\{(\psi^2\mathbb{U},\psi^2\mathbb{U})+(\mathbb{U},\mathbb{U})\},$$

$$|(\hat{\mathbb{U}}_1,\ B\cdot\mathbb{U}_2)|\leq\frac{1}{2}(\hat{\mathbb{U}}_1,\ \hat{\mathbb{U}}_1)\ +\ \frac{1}{2}(\mathbb{U}_2,\ \mathbb{U}_2),$$

so

$$\begin{aligned}J_k\ \geq\ h_x\cdot h_y\cdot&\sum_{i=1}^{\infty}\sum_{j=-\infty}^{\infty}\{(\hat{\mathbb{U}},\ A\cdot\hat{\mathbb{U}})+(\mathbb{U},\ A\cdot\mathbb{U})\}\ -\\ -\ \nu_C\cdot\Delta h_x\cdot&\sum_{i=1}^{\infty}\sum_{j=-\infty}^{\infty}\{(\hat{\mathbb{U}},\ \hat{\mathbb{U}})\ +\ (\mathbb{U},\ \mathbb{U})\}-\end{aligned}$$

$$- \Delta h_y \cdot \sum_{i=1}^{\infty} \sum_{j=-\infty}^{\infty} \{(\psi \hat{\mathrm{U}},\ \psi \hat{\mathrm{U}}) + \frac{1}{2}(\mathrm{U},\ \mathrm{U}) +$$

$$+ \frac{1}{2}(\psi^2 \mathrm{U},\ \psi^2 \mathrm{U})\} \ - \ \frac{1}{2}\Delta h_y \cdot \sum_{j=-\infty}^{\infty} \{(\hat{\mathrm{U}}_1,\ \hat{\mathrm{U}}_1)\ +\ (\mathrm{U}_2,\ \mathrm{U}_2)\} \ \geq$$

$$\geq\ h_x \cdot h_y \cdot \sum_{i=1}^{\infty} \sum_{j=-\infty}^{\infty} \{(\hat{\mathrm{U}},\ M \cdot \hat{\mathrm{U}})\ +\ (\mathrm{U},\ M \cdot \mathrm{U})\},$$

where $M\ =\ A\ -\ \nu_C \cdot r_y \cdot I_N\ -\ r_x \cdot I_N$. Thus, if the **condition of Courant-Friedrichs-Levi** is fulfilled (2.5) (see [1]), then $M\ >\ 0$. Consequently, $J_k\ >\ 0$ at $\mathrm{U},\ \hat{\mathrm{U}}\ \neq\ 0$.

Remark 2.1. Analisys of formulation of difference model (2.1)–(2.4) shows that not only additional boundary conditins but also additional initial conditions (i.e., the value of the vector U^1_{ij}) are required. To pose necessary initial conditions one can use any of two-layers difference schemes (for example, the one from the previous section).

3. Spectrum method for investigation of stability of difference schemes

We start now investigation of difference schemes for **mixed problem II**. We note that **problem II** differs in principal from **problem I** because boundary conditioons (0.8) are not dissipative for it. In this connection a more complicated investigation is to bbe carried out both for **mixed problem** and difference models of this problem. Before proceeding to construction of adequate difference models for **mixed problem II**, we will make brief acquaintance with the **spectrum method** and its application to investigation of difference models for **prolbem II** (for more detail see [9]).

Chosen difference scheme from section 1, we formulate a difference scheme for **problem II** (system (0.7) is considered simulteneously with the equation for S, see **introduction**):

$$\left\{A \cdot \tau + \frac{1}{2} r_x B \cdot \xi_0 + \frac{1}{2} r_y C \cdot \eta_0 - A \cdot (a \cdot \hat{\xi} + b \cdot \hat{\eta})\right\} \mathrm{U} = 0,$$

$$l = 1, 2, \cdots; \quad k, |j| = 0, 1, 2, \cdots; \tag{3.1}$$

$$\mathrm{U}^0_{lj} = \mathrm{U}_0(l \cdot h_x, j \cdot h_y), l, |j| = 0, 1, 2, \cdots; \tag{3.2}$$

$$\begin{cases} u^k_{0j} + d \cdot p^k_{0j} = 0, \quad S^k_{0j} + \cdot p^k_{0j} = 0, \\ \tau v^k_{0j} + \frac{r_y}{2} \omega \eta_0 v^k_{0j} - \lambda \frac{r_y}{2} \eta_0 p^k_{0j} - b \hat{\eta} v^k_{0j} = 0, \\ k, |j| = 0, 1, 2, \cdots. \end{cases} \tag{3.3}$$

Here $a, b > 0$ are aritrary as yet constants; $C = C_0 + \omega A$ (see **introduction**), $\mathbb{U} = \mathbb{U}_{ij}^k$. It is seen from (3.3) that an additional condition is requiered in order to determine $\hat{\mathbb{U}}_0$ In this connectioon there appears the question on stability of scheme (3.1) with account of boundary conditions (3.3) and also the additional boundary condition. Below it becomes clear that answer to this question is not evident.

First we will carry out the spectrum analysis of difference scheme of Cauchy problem (3.1), (3.2) (therewith $k, |l|, |j| = 0, 1, ...$). We seek particular solutions to (3.1) in the form:

$$\mathbb{U} = q^k \cdot \chi^l \cdot e^{i \cdot j \cdot \varphi_0} \cdot \mathbf{V}_0, \tag{3.4}$$

where $\mathbf{V}_0$ is a constant vector, φ_0 is a real parameter; χ, q are complex numbers, $i = \sqrt{-1}$.

Substituting (3.4) into (3.1), we obtain

$$\Big\{(q-1) \cdot I_4 + \frac{r}{2} A^{-1}[B \cdot \xi_0(\chi) + 2 \cdot i \cdot \sin\varphi_0 \cdot C] -$$

$$- \left[a \cdot \hat{\xi}(\chi) - 4\sin^2\frac{\varphi_0}{2} \cdot b\right] I_4 \Big\} \cdot \mathbf{V}_0 = 0, \tag{3.5}$$

where $\xi_0(\chi) = (\chi - \chi^{-1})$, $\hat{\xi}(\chi) = \chi - 2 + \chi^{-1}$, $r = r_x = r_y (h = h_x = h_y)$. $\chi = e^{i\psi_0}$, ψ_0 is a real parameter. Therewith system (3.5) has nontrivial solution $\mathbf{V}_0$ for $q = q(\varphi_0, \psi_0)$ such that the determinant of (3.5) turns into zero. The necessary condition on stability of (3.1) is validity of the following inequality for any φ_0, ψ_0

$$|q(\varphi_0, \psi_0)| \leq 1, \tag{3.6}$$

i.e., the spectrum of the difference Cauchy problem must lie in the unit circle (see also [9]). We check validity of (3.6) at $\varphi_0 = \psi_0$. In this case the determinant of (3.5) is equal to zero means that

$$q = q_1 = 1 - 4(a+b)\sin^2\frac{\varphi_0}{2} - r(\omega+1)\sin\varphi_0 \cdot i,$$

$$q = q_{2,3} = 1 - 4(a+b)\sin^2\frac{\varphi_0}{2} - r(\omega + 1 \pm \sqrt{2}\kappa)\sin\varphi_0 \cdot i,$$

$$\kappa = M^{-1}.$$

In view of $|q_{1,2,3}| \leq 1$ we come to the following restriction on a, b, r:

$$0 < a + b < \frac{1}{2},$$

$$r \leq \frac{\sqrt{2(a+b)}}{1 + \omega + \sqrt{2}\kappa}. \tag{3.7}$$

We consider now a one-dimensional variant of **mixed problem II** . In the domain $t > 0 x > 0$ we seek the solution to the acoustics equations system

$$\mathbf{A}_0 \mathbf{W}_t + \mathbf{B}_0 \mathbf{W}_x = 0, \tag{3.8}$$

which satisfies boundary conditions at $x = 0, t > 0$

$$u + d \cdot p = 0, \quad S + p = 0 \tag{3.9}$$

and initial data at $t = 0,\ x > 0$

$$\mathbf{W}(0, x) = \mathbf{W}_0(x). \tag{3.10}$$

Here

$$A_0 = \begin{pmatrix} M^2 & 0 & 0 \\ 0 & 1 & 0 \\ 0 & 0 & 1 \end{pmatrix}, \quad B_0 = \begin{pmatrix} M^2 & 1 & 0 \\ 1 & 1 & 0 \\ 0 & 0 & 1 \end{pmatrix},$$

$$\mathbf{W} = \begin{pmatrix} u \\ p \\ S \end{pmatrix}.$$

In this case difference model (3.1)–(3.3) will transorm into

$$\left\{ A_0 \cdot \tau + \frac{1}{2} r \cdot B_0 \cdot \xi_0 - a \cdot A_0 \cdot \hat{\xi} \right\} \cdot \mathbf{W} = 0.$$

$$l = 1, 2, \ldots, \quad k = 0, 1, 2, \ldots; \tag{3.11}$$

$$\mathbf{W}_l^0 = \mathbf{W}_0(l \cdot h), \quad l = 0, 1, 2, \ldots; \tag{3.12}$$

$$u_0^k + d \cdot p_0^k = 0, \quad s_0^k + p_0^k = 0, \quad k = 0, 1, 2, \ldots. \tag{3.13}$$

We carry out now the spectrum analysis for problem (3.11)–(3.13). We seek its solution of the form:

$$\mathbf{W} = q^k \cdot \chi^l \cdot \mathbf{Q}_0, \tag{3.14}$$

where

$$\mathbf{Q}_0 = \begin{pmatrix} u_0 \\ p_0 \\ S_0 \end{pmatrix} \text{ is constant vector.}$$

Substituting (3.14) into (3.11), we obtain

$$\begin{pmatrix} \sigma & g \cdot \kappa^2 & 0 \\ g & \sigma & 0 \\ 0 & 0 & \sigma \end{pmatrix} \cdot \mathbf{Q}_0 = 0, \tag{3.15}$$

where $\sigma = q - 1 - a \cdot \hat{\xi}(\chi) + g$, $g = \frac{1}{2} r \cdot \xi_0(\chi)$.

It follows from (3.15) that two types of solutions (3.14) are possible:

$$1)\quad S_0 \neq 0, \quad u_0 = p_0 = 0, \quad \sigma(\chi_0) = 0,$$

$$\mathbf{Q}_0 = \begin{pmatrix} 0 \\ 0 \\ S_0 \end{pmatrix}, \tag{3.16}$$

$$2)\quad \begin{cases} S_{1,2} = 0, \ \sigma(\chi_1) = -\kappa g(\chi_1), \ \sigma(\chi_2) = \kappa g(\chi_2), \\ \mathbf{Q}_1 = \begin{pmatrix} u_1 \\ p_1 \\ 0 \end{pmatrix}, \mathbf{Q}_2 = \begin{pmatrix} u_2 \\ p_2 \\ 0 \end{pmatrix}, \\ u_1 - \kappa p_1 = 0, \ u_2 + \kappa p_2 = 0. \end{cases} \tag{3.17}$$

While deducing (3.16), (3.17) we suppose $g(\chi_0) \neq 0$, $g(\chi_{1,2}) \neq 0$. With account of (3.16), (3.17), solutions to (3.11) are sought in the form:

$$\mathbf{W} = q^k \cdot [\chi_0^l \cdot \mathbf{Q}_0 + \chi_1^l \cdot \mathbf{Q}_1 + \chi_2^l \cdot \mathbf{Q}_2], \tag{3.18}$$

$|\chi_{0,1,2}| < 1, \chi_{1,2}$ are simple roots.

The following Lemma will be used in the sequel.

Lemma 3.1. *The roots of each of equations*

$$\sigma(\chi) = 0,$$

$$\sigma(\chi) = -\kappa g(\chi),$$

$$\sigma(\chi) = \kappa g(\chi)$$

satisfy the condition: at $|q| > 1$ *one of the roots lies inside the unit circle, another one lies outside of the unit circle.*

Proof. With the help of linear-fractional transformations

$$q = \frac{\rho + 1}{\rho - 1}, \quad \chi = \frac{\zeta + 1}{\zeta - 1}.$$

we rewritee $\sigma(\chi) = 0$ as

$$\zeta^2 + r(\rho - 1)\zeta - 1 - 2a(\rho - 1) = 0.$$

The roots of the equation are:

$$\zeta_{1,2} = \frac{-r(\rho - 1) \pm \sqrt{r^2(\rho - 1)^2 + 4 + 8a(\rho - 1)}}{2}.$$

Since

$$r^2(\rho - 1)^2 + 4 + 8a(\rho - 1) = [r(\rho - 1) + 4a/r]^2 + 4(1 - 4a^2/r^2),$$

$$\mathrm{Re}\left(\sqrt{r^2(\rho - 1)^2 + 4 + 8a(\rho - 1)}\right) \geq r\mathrm{Re}(\rho - 1) + 4a/r.$$

While dedicing the later, we suppose that a, r meet restrictions which were infered as the result of the spectrum analysis of the one-dimensional Cauchy problem:

$$0 < a < \frac{1}{2}, \quad r \leq \frac{\sqrt{2a}}{k_1}, \quad k_1 = 1 + \kappa. \tag{3.19}$$

Indeed, by (3.19),

$$\frac{4a}{r} - r > 0, \quad 1 - \frac{4a^2}{r^2} \geq 0.$$

Consequently, at $|q| > 1(\text{Re}\rho > 0)$:

$$\text{Re}\zeta_1 > 0 \quad (\text{i.e.}|\chi_1| > 1),$$

$$\text{Re}\zeta_2 < 0 \quad (\text{i.e.}|\chi_2| < 1),$$

Equations

$$\sigma(\chi) = -\kappa g(\chi), \quad \sigma(\chi) = \kappa g(\chi).$$

are analysed by analogy.

Remark 3.1. At $q = 1$ the roots of equations are: for $\sigma(\chi) = 0$

$$\chi_1 = 1, \quad \chi_2 = \frac{2a + r}{2a - r};$$

for $\sigma(\chi) = -\kappa g(\chi)$

$$\chi_1 = 1, \quad \chi_2 = \frac{2Ma + r(1 + M)}{2Ma - r(1 + M)};$$

for $\sigma(\chi) = \kappa g(\chi)$

$$\chi_1 = 1, \quad \chi_2 = \frac{2Ma - r(1 - M)}{2Ma + r(1 - M)} = \hat{\chi}.$$

By virtue of **Lemma 3.1**, as χ_0, χ_1, χ_2 from (3.18) we take

$$\begin{cases} \chi_0 = \dfrac{2a + q - 1 - \sqrt{r^2 + (q - 1)(4a + q - 1)}}{2a - r}, \\ \\ \chi_1 = \dfrac{2a + q - 1 - \sqrt{r^2 k_1^2 + (q - 1)(4a + q - 1)}}{2a - rk_1}, \\ \\ \chi_2 = \dfrac{2a + q - 1 - \sqrt{r^2 k_2^2 + (q - 1)(4a + q - 1)}}{2a + rk_2}, \end{cases} \tag{3.20}$$

where $k_2 = \kappa - 1$.

We already know that an additional condition is necessary in order to determine $\hat{\mathbf{W}}_0$. With this purpose in a view, we write first a differential relation, a consequence of system (3.8),

$$\mathbf{W}_t + \mathbf{W}_x + \kappa^2 p_x + \hat{\beta} u_x = 0, \tag{3.21}$$

$$\mathbf{W} = u + \hat{\beta} p,$$

where $\hat{\beta}$ is a constant , and then an approximation of (3.21) at $x = 0$:

$$\tau \mathbf{W}_0^k + (1 + \hat{\beta})\gamma u_0^k + (\kappa^2 + \hat{\beta})\mu p_0^k = 0. \tag{3.22}$$

Here $\gamma(\psi), \mu(\psi)$ are difference operators approximating the operator $\Delta \cdot \dfrac{\partial}{\partial x}$ at $x = 0$ (for example, $\gamma(\psi) = r(\psi - 1)$ an so on). With account of (3.17), (3.22), we obtain a linear algebraic system to determine u_1, p_1, u_2, p_2. This system must have nontrivial solution i.e., the determinant of the system is not identical zero. From this condition we derive the determinant relation for q:

$$(d - \kappa)(a_2 + \kappa a_0) = (d + \kappa)(a_3 - \kappa a_1), \tag{3.23}$$

where

$$a_0 = q - 1 + (1 + \hat{\beta}) \cdot \gamma(\chi_1),$$

$$a_1 = q - 1 + (1 + \hat{\beta}) \cdot \gamma(\chi_2),$$

$$a_2 = \hat{\beta}(q - 1) + (\kappa^2 + \hat{\beta}) \cdot \mu(\chi_1),$$

$$a_3 = \hat{\beta}(q - 1) + (\kappa^2 + \hat{\beta}) \cdot \mu(\chi_2).$$

Let $\hat{\beta} = -\kappa$, $\gamma = \mu = r(\psi - 1)$. Then (3.23) transforms into ($d \neq -\kappa$):

$$q - 1 = k_2 r(\chi_2 - 1) \tag{3.23$'$}$$

However, relation (3.23′) does not has solutions (under $\dfrac{4a^2}{r^2 k_2^2} \neq 1$). Consequently, in the one-dimensional case it is possibble to find the boundary condition such that (3.11)–(3.13) does not admit existence of the Hadamard example, in other words, it does not have solution of the form (3.18) with $|q| > 1$.

Now turn back to (3.1)–(3.3). We supplement (3.3) with one more condition. For this purpose we write down the differential relation

$$w_t + w_x + \omega w_y + \beta_2 u_x + \beta_0 \kappa^2 p_x + \beta_2 v_y + \beta_1 \kappa^2 p_y = 0, \tag{3.24}$$

$$w = \beta_0 \cdot u + \beta_1 \cdot v + \beta_2 \cdot p,$$

which is a consequence of (0.7), and an approximation of (3.24) at $x = 0$:

$$\tau w_{0j}^k + (\beta_0 + \beta_2) \cdot \gamma u_{0j}^k + \beta_1 \delta v_{0j}^k + (\beta_0 \kappa^2 + \beta_2) \cdot \mu p_{0j}^k + \frac{1}{2} r \beta_0 \omega \eta_0 u_{0j}^k +$$

$$+\frac{1}{2}r(\omega\beta_1+\beta_2)\cdot\eta_0 v_{0j}^k+\frac{1}{2}r(\beta_1\kappa^2+\omega\beta_2)\cdot\eta_0 p_{0j}^k-b\hat{\eta}w_{0j}^k=0. \qquad (3.25)$$

Here $\beta_{0,1,2}$ are constants; $\gamma(\psi)$, $\delta(\psi)$, $\mu(\psi)$ are diffeence operators approximating the operator $\Delta\cdot\frac{\partial}{\partial x}$ at $x=0$ (for example, $\gamma=\frac{r}{2}(4\psi-\psi^2-3)$ and so on); $r=r_x=r_y$ $(h=h_x=h_y)$.

The following is true

Theorem 3.1. *Difference scheme (3.1) with boundary conditions (3.3) and (3.25) is unstable for any β_0,β_1,β_2 and for any $\gamma(\psi)$, $\delta(\psi)$, $\mu(\psi)$, approximating the differential operator $\Delta\cdot\frac{\partial}{\partial x}$ at $x=0$.*

Proof. Rewrite (3.5) as a follows:

$$\begin{pmatrix} \sigma & 0 & g\kappa^2 & 0 \\ 0 & \sigma & \nu\kappa^2 & 0 \\ g & \nu & \sigma & 0 \\ 0 & 0 & 0 & \sigma \end{pmatrix}\cdot\mathbf{V}_0=0, \qquad (3.5')$$

where

$$\sigma=q-1-a\cdot\hat{\xi}(\chi)+g+4b\cdot\sin^2(\frac{\varphi_0}{2})+\omega\nu,$$

$$g=\frac{1}{2}r\cdot\xi_0(\chi),\nu=i\cdot r\cdot\sin\varphi_0.$$

Solution to (3.1) is sought in the form :

$$\mathbf{U}=q^k\cdot\exp\{j\varphi_0 i\}[\chi_0^l\cdot\mathbf{V}_0+\chi_1^l\cdot\mathbf{V}_1+\chi_2^l\cdot\mathbf{V}_2], \qquad (3.26)$$

where φ_0 is a real parameter; χ_0 is a root of $\sigma(\chi)=0$; χ_1, χ_2 are roots of $M^2\sigma^2(\chi)=g^2(\chi)+\nu^2$, and χ_1, χ_2 are simple roots, and $|\chi_{0,1,2}|<1$;

$$\mathbf{V}_0=\begin{pmatrix} u_0 \\ v_0 \\ 0 \\ S_0 \end{pmatrix},\quad \mathbf{V}_1=\begin{pmatrix} u_1 \\ v_1 \\ p_1 \\ 0 \end{pmatrix},\quad \mathbf{V}_2=\begin{pmatrix} u_2 \\ v_2 \\ p_2 \\ 0 \end{pmatrix},$$

relations on $\mathbf{V}_{0,1,2}$ are (see (3.5′)):

$$\begin{cases} g(\chi_0)\cdot u_0+\nu v_0=0, \\ \sigma(\chi_1)\cdot u_1+g(\chi_1)\kappa^2\cdot p_1=0, \\ \sigma(\chi_1)\cdot v_1+\nu\kappa^2\cdot p_1=0, \\ \sigma(\chi_2)\cdot u_2+g(\chi_2)\kappa^2\cdot p_2=0, \\ \sigma(\chi_2)\cdot v_1+\nu\kappa^2\cdot p_2=0. \end{cases} \qquad (3.27)$$

Since at $\varphi_0=0$ and $q=1$: $\chi_0=1,\chi_1=1,\chi_2=\hat{\chi}$ (see **Remark 3.1**) we will investigate the formulated difference model on "long waves" under assumption

$\varphi_0 = \varepsilon\hat{\varphi}$, where ε is a small parameter, $\hat{\varphi} = O(1)$. We will consider the relations

$$\begin{cases} q - 1 = \varepsilon\hat{q} + O(\varepsilon^2), \\ \chi_0 - 1 = \varepsilon\hat{\chi_0} + O(\varepsilon^2), \\ \chi_1 - 1 = \varepsilon\hat{\chi_1} + O(\varepsilon^2), \\ \chi_2 - \hat{\chi} = \varepsilon\hat{\chi_2} + O(\varepsilon^2). \end{cases} \tag{3.28}$$

to be valid.

Relations (3.27), boundary conditions (3.3), (3.25) form a linear algebraic system to determine u_0, v_0, u_1, v_1, p_1, u_2, v_2, p_2. The determinant condition on q is derived from requirement on the determinant of this system to be equal to zero:

$$\begin{gathered} \hat{\sigma}\cdot\sigma(\chi_1)\cdot\sigma(\chi_2)\cdot g(\chi_0)\cdot d(a_6 - a_7) - \sigma(\chi_1)\cdot\sigma(\chi_2)\cdot\lambda\cdot\nu^2\cdot(a_6 - a_7)+ \\ +\sigma(\chi_2)\cdot g(\chi_0)\cdot g(\chi_1)\cdot\kappa^2(a_7\hat{\sigma} - a_1\hat{\sigma}d + a_3\lambda\nu)+ \\ +g(\chi_1)\cdot\sigma(\chi_2)\cdot\lambda\cdot\kappa^2\cdot\nu^2\cdot(a_1 - a_0) + \sigma(\chi_2)\cdot\nu^2\cdot\kappa^2(a_7\hat{\sigma} - a_0\hat{\sigma}d + a_4\lambda\nu)- \\ -\sigma(\chi_1)\cdot g(\chi_0)\cdot g(\chi_2)\cdot\kappa^2(a_6\hat{\sigma} - a_2\hat{\sigma}d + a_3\lambda\nu)- \\ -g(\chi_2)\cdot\sigma(\chi_1)\cdot\lambda\cdot\kappa^2\cdot\nu^2\cdot(a_2 - a_0) + \hat{\sigma}\cdot g(\chi_1)\cdot g(\chi_2)\cdot g(\chi_0)\cdot\kappa^4(a_1 - a_2)+ \\ +\hat{\sigma}\cdot g(\chi_2)\cdot\nu^2\cdot\kappa^4(a_0 - a_2) - \sigma(\chi_1)\cdot\nu^2\cdot\kappa^2(a_6\hat{\sigma}- \;; a_0\hat{\sigma}d + a_5\lambda\nu)- \\ -\hat{\sigma}\cdot g(\chi_1)\cdot\nu^2\cdot\kappa^4(a_0 - a_1) + \sigma(\chi_2)\cdot\nu\cdot\kappa^2\cdot d\cdot g(\chi_0)\cdot\hat{\sigma}(a_3 - a_4)+ \\ +g(\chi_2)\cdot\nu\cdot\kappa^4\cdot g(\chi_0)\cdot\hat{\sigma}(a_4 - a_3) + \sigma(\chi_1)\cdot\nu\cdot\kappa^2\cdot d\cdot g(\chi_0)\cdot\hat{\sigma}(a_5 - a_3)- \\ -g(\chi_1)\cdot\nu\cdot\kappa^4\cdot g(\chi_0)\cdot\hat{\sigma}(a_5 - a_3) - \nu^3\cdot\kappa^4\cdot\hat{\sigma}(a_5 - a_4) = 0. \end{gathered} \tag{3.29}$$

Here

$$\begin{aligned} \hat{\sigma} &= q - 1 + \omega\nu + 4b\cdot\sin^2\frac{\varphi_0}{2}, \\ a_0 &= \beta_0\cdot\hat{\sigma} + (\beta_0 + \beta_2)\cdot\gamma(\chi_0), \\ a_1 &= \beta_0\cdot\hat{\sigma} + (\beta_0 + \beta_2)\cdot\gamma(\chi_1), \\ a_2 &= \beta_0\cdot\hat{\sigma} + (\beta_0 + \beta_2)\cdot\gamma(\chi_2), \\ a_3 &= \beta_1\cdot\hat{\sigma} + \beta_2\cdot\nu + \beta_1\cdot\delta(\chi_0), \\ a_4 &= \beta_1\cdot\hat{\sigma} + \beta_2\cdot\nu + \beta_1\cdot\delta(\chi_1), \\ a_5 &= \beta_1\cdot\hat{\sigma} + \beta_2\cdot\nu + \beta_1\cdot\delta(\chi_2), \\ a_6 &= \beta_2\cdot\hat{\sigma} + \beta_1\cdot\nu\kappa^2 + (\beta_0\kappa^2 + \beta_2)\cdot\mu(\chi_1), \\ a_7 &= \beta_2\cdot\hat{\sigma} + \beta_1\cdot\nu\kappa^2 + (\beta_0\kappa^2 + \beta_2)\cdot\mu(\chi_2). \end{aligned}$$

We substitute (3.28) into (3.29) and into

$$\sigma(\chi_0) = 0,$$

$$M^2\sigma^2(\chi_{1,2}) = g^2(\chi_{1,2}) + \nu^2,$$

and obtain

$$\hat{\chi}_0 = -\frac{\bar{q}}{r}, \quad \hat{q} = \bar{q} - ir\omega\hat{\varphi},$$

$$\hat{\chi}_1 = \hat{\varphi}\frac{1+z^2}{2z}, \quad \bar{q} = r\hat{\varphi}\frac{\kappa(1-z^2)-1-z^2}{2z},$$

$$(Mm_1 - m_2)[(z^2k_1 - k_2)^2(dM(1-z^2)-1-z^2)+ \\ +4z^2(z^2k_1-k_2) - 4z^2(1-z^2)\cdot\lambda M] = 0, \tag{3.30}$$

where

$$m_1 = (\kappa^2\beta_0 + \beta_2)\cdot\mu(\hat{\chi}), \quad m_2 = (\beta_0+\beta_2)\cdot\gamma(\hat{\chi}).$$

While deriving (3.30) we assume that

$$\gamma(\chi_0) = r\hat{\chi}_0\varepsilon + O(\varepsilon^2), \quad \delta(\chi_0) = r\hat{\chi}_0\varepsilon + O(\varepsilon^2),$$

$$\gamma(\chi_1) = r\hat{\chi}_1\varepsilon + O(\varepsilon^2), \quad \delta(\chi_1) = r\hat{\chi}_1\varepsilon + O(\varepsilon^2),$$

$$\mu(\chi_1) = r\hat{\chi}_1\varepsilon + O(\varepsilon^2). \tag{3.31}$$

It becomes clear for any reasonable approximations of $\Delta\cdot\dfrac{\partial}{\partial x}$ at $x=0$ (3.31) are valid. We also assume that

$$\gamma(\chi_2) = \gamma(\hat{\chi}) + O(\varepsilon), \quad \delta(\chi_2) = \delta(\hat{\chi}) + O(\varepsilon), \quad \mu(\chi_2) = \mu(\hat{\chi}) + O(\varepsilon),$$

therewith $\gamma(\hat{\chi}), \mu(\hat{\chi}) \neq 0$.

It is easy to see that the last relation from (3.30) has the root $z^2 = 1$. Therewith

$$\bar{q} = -\frac{r\hat{\varphi}}{z}, \quad \hat{\chi}_0 = \frac{\hat{\varphi}}{z}, \quad \hat{\chi}_1 = \frac{\hat{\varphi}}{z}.$$

Let $z = -\text{sign}(\hat{\varphi})$. Then

$$\text{Re}\bar{q} = \text{Re}\hat{q} > 0, \quad \text{Re}\hat{\chi}_0 = \text{Re}\hat{\chi}_1 < 0$$

and

$$|\chi_0|^2 = 1 + 2\varepsilon\text{Re}\hat{\chi}_0 + O(\varepsilon^2) < 1,$$

$$|\chi_1|^2 = 1 + 2\varepsilon\text{Re}\hat{\chi}_1 + O(\varepsilon^2) < 1,$$

$$|\chi_2|^2 = \hat{\chi}^2 + 2\varepsilon\hat{\chi}\text{Re}\hat{\chi}_2 + O(\varepsilon^2) < 1,$$

$$|q|^2 = 1 + 2\varepsilon\text{Re}\hat{q} + O(\varepsilon^2) > 1.$$

The last inequalities mean that we constructed the Hadamard example which confirms the instability of (3.1) (with account of the boundary conditions).

Remark 3.2. Theorem 3.1 holds true also for the cases when we either use another approximations of $\dfrac{\partial}{\partial y}$ in (3.1)–(3.3), (3.25) or add terms of the form $\hat{b}\hat{\eta}\mathrm{U}$, $\hat{b}\hat{\eta}\xi\mathrm{U}$ and so on ($\hat{b}$ is a constant) into (3.1), into the third condition from

(3.3), into (3.25). This fact is evident since the difference Hadamard example was constructed by the first approximation, i.e., the first terms from (3.28) were used (see Proof of **Theorem 3.1**).

An analogous conclusion is true if we add a term of the type $\hat{a}\cdot\gamma^2(\psi)w_{0j}^k$, $\hat{a}\cdot\gamma(\psi)\delta(\psi)w_{0j}^k$ ($\hat{a}$ is a constant) and so on into additional boundary condition (3.25) or a term of the type $\hat{a}\cdot\gamma^2(\psi)v_{0j}^k$, $\hat{a}\cdot\gamma(\psi)\delta(\psi)v_{0j}^k$ and so on into the third boundary condition (3.3).

Remark 3.3. The spectrum analysis carried out in this section shows that stability depends on a type of problem (is it the difference Cauchy problem or difference boundary-value problem) and its dimension. In this connection recommendation obtained for the difference Cauchy probblem or one-dimensional difference boundary-value probblem can not be automatically applied to multi-dimensional cases of the difference boundary-value problem.

Remark 3.4. This section is written on the basis of papers [10], [11].

4. Influence of boundary conditions on stability of a certain difference splitting scheme

In this section we will illustrate once more **Remark 3.3** concerning the fact, that conclusions obtained while investigating stability of the difference Cauchy problem are not automatically valid for the case of the difference boundary-value problem, which often is wrongly interpreted in usual calculation practice.

As an example we consider a difference model for **mixed problem II** which is based on splitting by space variables [12]:

$$\left\{A\tau^* + \frac{1}{2}r_y C L_t^* \eta_0\right\}\mathrm{U} = 0, \quad k, l, |j| = 0, 1, \ldots, \tag{4.1}$$

$$\{A\tau^* + r_x B_1 L_t^* \bar{\xi} - r_x B_2 L_t^* \xi\}\tilde{\mathrm{U}} = 0, \quad k, |j| = 0, 1, \ldots, \quad l = 1, 2, \ldots, \tag{4.2}$$

$$\mathrm{U}_{lj}^0 = \mathrm{U}_0(l\cdot h_x, j\cdot h_y), \quad l, |j| = 0, 1, 2, \ldots; \tag{4.3}$$

$$\left\{\begin{array}{l} u_{0j}^k + d\cdot p_{0j}^k = 0, \quad S_{0j}^k + \cdot p_{0j}^k = 0, \\ \tau v_{0j}^k + r_y\omega L_t\bar{\eta}v_{0j}^k - \lambda r_y L_t\bar{\eta}p_{0j}^k = 0, \\ k, |j| = 0, 1, 2, \ldots. \end{array}\right. \tag{4.4}$$

Here $\tau^* = \varphi^{\frac{1}{2}} - 1, L_t^* = \alpha\varphi^{\frac{1}{2}} + \delta, L_t = \alpha_0\varphi + \delta_0, \alpha + \delta = 1, \alpha_0 + \delta_0 = 1,$

$\alpha, \alpha_0, \delta, \delta_0 \geq 0$ are constants, $\varphi^{\frac{1}{2}} U_{lj}^k = U_{lj}^{k+\frac{1}{2}} = \tilde{U}$ $B = B_1 - B_2$,

$$B_1 = \begin{pmatrix} \frac{k_1}{2} & \frac{Mk_1}{2} & 0 & 0 \\ \frac{Mk_1}{2} & \frac{M^2 k_1}{2} & 0 & 0 \\ 0 & 0 & M^2 & 0 \\ 0 & 0 & 0 & 1 \end{pmatrix},$$

$$B_2 = \begin{pmatrix} \frac{k_2}{2} & -\frac{Mk_2}{2} & 0 & 0 \\ -\frac{Mk_2}{2} & \frac{M^2 k_2}{2} & 0 & 0 \\ 0 & 0 & 0 & 0 \\ 0 & 0 & 0 & 0 \end{pmatrix}.$$

Calculations with the help of (4.1)–(4.4) are carried out as follows: first, components of the vector $\mathbb{U}$ are calculated on the time layer $k + \frac{1}{2}$ with the help of (4.1) (therewith only initial data (4.3) are taken at this stage), then at the second half-step with the help of (4.2), (4.4) components of $\hat{\mathbb{U}}$ are calculated at the time layer $k+1$.

First we study the difference Cauchy problem (4.1)–(4.3) (here k, $|l|$, $|j| = 0, 1, \ldots$). We prove stability of this problem by means of the energy method [13]. With this purpose in a view we multiply (4.1) scalarly by $2L_t^* \mathbb{U}$:

$$\left(A\tau^* \mathbb{U} + \frac{1}{2} r_y C L_t^* \eta_0 \mathbb{U}, 2L_t^* \mathbb{U} \right) = \tau^*(\mathbb{U}, A\mathbb{U}) +$$

$$+(\alpha - \delta)(\tau^* \mathbb{U}, A\tau^* \mathbb{U}) + \frac{1}{2} r_y \{\eta(L_t^* \mathbb{U}, C L_t^* \mathbb{U}) +$$

$$+\bar{\eta}(L_t^* \mathbb{U}, C L_t^* \mathbb{U}) - \bar{\eta}(\eta L_t^* \mathbb{U}, \eta C L_t^* \mathbb{U})\} = 0.$$

Let $\alpha > \delta$. Then the last relation leads to

$$||\tilde{\mathbb{U}}||_A^2 \leq ||\mathbb{U}||_A^2, \tag{4.5}$$

where

$$||\mathbb{U}||_A^2 = h_x h_y \sum_{l,j=-\infty}^{\infty} (\mathbb{U}, A\mathbb{U}).$$

Then we multiply (4.2) scalarly by $2L_t^*\mathbb{U}$:

$$(A\tau^*\mathbb{U} + r_x B_1 L_t^* \bar{\xi}\tilde{\mathbb{U}} - r_x B_2 L_t^* \xi \mathbb{U}, 2L_t^*\tilde{\mathbb{U}}) =$$

$$= \tau^*(\tilde{\mathbb{U}}, A\tilde{\mathbb{U}}) + (\alpha - \delta)(\tau^*\tilde{\mathbb{U}}, A\tau^*\tilde{\mathbb{U}}) + r_x\bar{\xi}(B_1 L_t^*\tilde{\mathbb{U}}, L_t^*\tilde{\mathbb{U}}) +$$

$$+ r_x(B_1\bar{\xi}L_t^*\tilde{\mathbb{U}}, \bar{\xi}L_t^*\tilde{\mathbb{U}}) - r_x\xi(B_2 L_t^*\tilde{\mathbb{U}}, L_t^*\tilde{\mathbb{U}}) + r_x(B_2\xi L_t^*\tilde{\mathbb{U}}, \xi L_t^*\tilde{\mathbb{U}}) = 0.$$

Since $B_{1,2} \geq 0$ we obtain ($\alpha > \delta$!)

$$||\hat{\mathbb{U}}||_A^2 \leq ||\tilde{\mathbb{U}}||_A^2. \tag{4.6}$$

(4.5), (4.6) easily leads to the desired energy inequality

$$||\hat{\mathbb{U}}||_A^2 \leq ||\mathbb{U}||_A^2, \tag{4.7}$$

and, consequently to stability of difference scheme of splitting (4.1), (4.2) for the Cauchy problem. The sufficient condition on stability is fulfilment of the inequality $\alpha > \delta$.

In order to obtain necessary conditions on stability of scheme (4.1), (4.2) we will analyse its spectrum. We seek special solutions to (4.1), (4.2):

$$\mathbb{U} = q^k \cdot \chi^l \cdot e^{ij\varphi_0} \cdot V_0, \tag{4.8}$$

where V_0 is a constant vector, φ_0 is a real parameter , q, χ are complex. Substituting (4.8) into (4.1) and into (4.2), we come to

$$\begin{gathered}\{[A + \alpha g B_1 - \alpha g \chi B_2]q + \\ + [-A + \delta g B_1 - \delta g \chi B_2] A^{-1/2} T^{-1} \hat{N} T A^{1/2}\} V_0 = 0,\end{gathered} \tag{4.9}$$

where $g = r_x(1 - \chi^{-1})$,

$$T = \begin{pmatrix} 1 & 0 & 1 & 0 \\ 0 & 1 & 0 & 0 \\ 1 & 0 & -1 & 0 \\ 0 & 0 & 0 & 1 \end{pmatrix}, \quad T^{-1} = \begin{pmatrix} \frac{1}{2} & 0 & \frac{1}{2} & 0 \\ 0 & 1 & 0 & 0 \\ \frac{1}{2} & 0 & -\frac{1}{2} & 0 \\ 0 & 0 & 0 & 1 \end{pmatrix},$$

$$\hat{N} = \begin{pmatrix} b_1 + b_3 & 0 & 0 & 0 \\ 0 & b_2 & 0 & 0 \\ 0 & 0 & b_1 - b_3 & 0 \\ 0 & 0 & 0 & b_2 \end{pmatrix},$$

$$b_1 = \frac{1 + \nu\omega(\alpha - \delta) + \nu^2\alpha\delta(\kappa^2 - \omega^2)}{1 + 2\nu\omega\alpha - \nu^2\alpha^2(\kappa^2 - \omega^2)},$$

$$b_2 = \frac{1 - \nu\omega\delta}{1 + \nu\omega\delta}, \nu = ir_y \sin\varphi_0,$$

$$b_3 = -\frac{\nu\kappa}{1 + 2\nu\omega\alpha - \nu^2\alpha^2(\kappa^2 - \omega^2)}.$$

For the difference Cauchy problem $\chi = e^{i\psi_0}$, ψ_0 is real. System (4.9) has a nontrivial solution V_0 only for $q = q(\varphi_0, \psi_0)$ such that the determinant of system (4.7) turns into zero. The necessary condition for model (4.1), (4.2) to be stable is validity of the following inequality for any φ_0, ψ_0:

$$|q(\varphi_0, \psi_0)| \leq 1, \tag{4.10}$$

i.e., the spectrum of the problem lies in the unit disk.

Because of awkwardness of expressions for the roots q we will restrict ourself by the case of "long waves": assuming that $\varphi_0 = \varepsilon\hat{\varphi}$, $\hat{\varphi} = O(1)$, ε is a small parameter, we check validity of (4.10) at $\psi_0 = \varphi_0$. In this case

$$q = 1 + \varepsilon\hat{q} + \varepsilon^2\check{q} + O(\varepsilon^2). \tag{4.11}$$

And (4.10) holds true if Re $\hat{q} < 0$. For the case Re $\hat{q} = 0$, (4.10) is equivalent to the following

$$2\text{Re}\check{q} + (\text{Jm}\hat{q})^2 \leq 0. \tag{4.12}$$

After simple calculations, with the help of (4.11), we determine the roots q:

$$\hat{q}_1 = -i\hat{\varphi}\bar{r}, \bar{r} = (r_x + \omega r_y),$$

$$\check{q}_1 = -\hat{\varphi}^2\left[\alpha\bar{r}^2 - (\alpha - \delta)\omega r_x r_y + \frac{1}{2}r_x\right],$$

$$\hat{q}_{2,3,4} = i\hat{\varphi}[\sigma\kappa\hat{r} - \bar{r}], \quad \sigma = 0, \pm 1, \quad \hat{r} = \sqrt{r_x^2 + r_y^2},$$

$$\check{q}_2 = \hat{\varphi}^2\left[-\frac{\frac{1}{2}\kappa r_x r_y^2 + \frac{1}{2}r_x^3 + (\alpha - \delta)\kappa^2 r_x^2 r_y^2}{\hat{r}^2} - \alpha r_x\bar{r} - \alpha\omega^2 r_y^2 - \delta\omega r_x r_y\right],$$

$$\check{q}_{3,4} = \hat{\varphi}^2\left[\frac{\frac{1}{2}\kappa r_x r_y^2 + \frac{1}{2}r_x^3 + (\alpha - \delta)\kappa^2 r_x^2 r_y^2}{2\hat{r}^2} + \right.$$

$$+\sigma\frac{r_x^2(\alpha\kappa r_x + \kappa\omega r_y + \frac{1}{2}) + \kappa r_y^2(2\alpha\omega r_y + \delta r_x)}{\hat{r}} -$$

$$\left. -\alpha\kappa^2\hat{r}^2 - r_x\frac{\kappa + \frac{1}{2}}{2} + \alpha r_x(\sigma\kappa\hat{r}^2 - \bar{r}) - \alpha\omega^2 r_y - \delta\omega r_x r_y\right],$$

$$\sigma = \pm 1.$$

Sustituting expressions for Re $\hat{q}_j$, Im $\check{q}_j (j = 1, 2, 3, 4)$ into (4.12) we arrive at the inequalities: $j = 1$:

$$(\alpha - \delta)(r_x^2 + \omega^2 r_y^2) + r_x \geq 0,$$

$j = 2$:

$$(\alpha - \delta)(r_x^2 + \omega^2 r_y^2) + \frac{\kappa r_x r_y^2 + r_x^3 + 2(\alpha - \delta)\kappa^2 r_x^2 r_y^2}{r_x^2 + r_y^2} \geq 0,$$

$j = 3, 4 (\sigma = \pm 1)$:

$$2\sigma\kappa\hat{r}^2[r_x^3(\frac{3\alpha}{2} - 1) - \frac{\alpha r_x r_y^2}{2} + (\alpha - \delta)\omega r_y^3 + \frac{M r_x^2}{2}] \leq$$

$$\leq (\alpha - \delta)[r_x^4(1 + \kappa^2) + r_x^2 r_y^2(1 + \kappa^2 + \omega^2) + r_y^4(\kappa^2 + \omega^2)] + \kappa r_x^3 + r_x r_y^2 \frac{\kappa + 1}{2}.$$

It is easy to show that at $\alpha > \delta$ the inequalities are fulfilled , i.e., the necessary condition on stability of (4.10) is fulfilled.

Let us turn back to mixed problem (4.1)–(4.4). Algebraic system (4.9) can be rewritten as follows:

$$A^{-1/2} \cdot F \cdot A^{1/2} \cdot \mathbf{V}_0 = 0, \qquad (4.9')$$

where

$$F = \begin{pmatrix} f_1 & f_2 & f_3 & 0 \\ f_4 & f_5 & f_6 & 0 \\ f_7 & 0 & f_8 & 0 \\ 0 & 0 & 0 & f_9 \end{pmatrix},$$

$$f_1 = q[1 + \alpha g R_1] + b_1[\delta g R_1 - 1],$$

$$f_2 = g R_2(\alpha q + \delta b_2),$$

$$f_3 = b_3[\delta g R_1 - 1],$$

$$f_4 = g R_2(\alpha q + \delta b_1),$$

$$f_5 = q[1 + \alpha g R_1] + b_2[\delta g R_1 - 1],$$

$$f_6 = b_3 \delta g R_2,$$

$$f_7 = b_3(\delta g - 1),$$

$$f_8 = q[1 + \alpha g] + b_1[\delta g - 1],$$

$$f_9 = q[1 + \alpha g] + b_2[\delta g - 1],$$

$$R_1 = \frac{k_1 - \chi k_2}{2} R_2 = \frac{k_1 + \chi k_2}{2}$$

$\mathbf{V}_0 = (p_0, u_0, v_0, s_0)^*$ is the sought vector.

Existence of the solution of the type (4.8) such that $|\chi| \leq 1$, $|q| > 1$ means existence of the Hadamard examples, and by this, instability of problem (4.1)–(4.4).

From the condition

$$\det F = 0 \qquad (4.13)$$

it follows that two types of the solutions are expected.

1) $S_0 \neq 0$, $f_9 = 0$,

$$\mathbf{Q}_0 = \begin{pmatrix} 0 \\ 0 \\ 0 \\ S_0 \end{pmatrix}, \quad \chi_0 = \frac{r_x(\alpha q + \delta b_2)}{q(\alpha r_x + 1) + b_2(\delta r_x - 1)} = 1 + \varepsilon\hat{\chi}_0 + O(\varepsilon^2),$$

$$\hat{\chi}_0 = -\frac{\bar{q}}{r_x}, \quad \bar{q} = \hat{q} + i\hat{\varphi} r_y \omega;$$

2) $S_0 = 0$, $f_9 \neq 0$
Presenting f_j, $j = \overline{1,8}$ in the form

$$f_j = \varepsilon \hat{f}_j + \varepsilon^2 \check{f}_j + O(\varepsilon^2),$$

from (4.13) in the first approximation we obtain the equation:

$$\hat{f}_1^2 \cdot (\hat{f}_1^2 - \hat{f}_2^2 - \hat{f}_3^2) = 0,$$

With account of (4.14), we obtain expressions for two more roots $\chi_{1,2}$:

$$\chi_1 = 1 + \varepsilon \hat{\chi}_1 + O(\varepsilon^2),$$

$$\hat{\chi}_1 = -\frac{\bar{q}}{r_x}, \chi_2 = 1 + \varepsilon \hat{\chi}_2 + O(\varepsilon^2),$$

$$\hat{\chi}_2 = \frac{\bar{q} - \sqrt{\bar{q}^2 \kappa^2 + (\kappa^2 - 1)\hat{\varphi}_0^2 r_y^2 \kappa^2}}{r_x(\kappa^2 - 1)},$$

$$\mathbf{Q}_1 = \begin{pmatrix} p_1 \\ u_1 \\ v_1 \\ 0 \end{pmatrix}, \quad \mathbf{Q}_2 = \begin{pmatrix} p_2 \\ u_2 \\ v_2 \\ 0 \end{pmatrix}$$

The third root χ_3 is omitted since $|q| > 1$: $|\chi_3| > 1$.

By virtue of (4.9′), $p_1, u_1, v_1, p_2, u_2, v_2$, are connected by the relations:

$$\begin{cases} f_1(\chi_1) \cdot p_1 + M \cdot f_2(\chi_1) \cdot u_1 + M \cdot f_3(\chi_1) \cdot v_1 = 0, \\ f_7(\chi_1) \cdot p_1 + M \cdot f_8(\chi_1) \cdot v_1 = 0, \\ f_1(\chi_2) \cdot p_2 + M \cdot f_2(\chi_2) \cdot u_2 + M \cdot f_3(\chi_2) \cdot v_2 = 0, \\ f_7(\chi_2) \cdot p_2 + M \cdot f_8(\chi_2) \cdot v_2 = 0. \end{cases} \tag{4.15}$$

Thus, the desired solution has the form:

$$\mathbf{U} = q^k \cdot e^{ij\varphi_0}[\chi_0^l \cdot \mathbf{Q}_0 + \chi_1^l \cdot \mathbf{Q}_1 + \chi_2^l \cdot \mathbf{Q}_2]. \tag{4.16}$$

(4.15) together with boundary conditions (4.4) serves as a system to determine $p_1, u_1, v_1, p_2, u_2, v_2$. The system must have a nontrivial soltution, i.e., the determinator of this system must equal zero. This consition yields the determinant relation on q. Its first approximation is

$$d(\hat{q} - i\hat{\varphi} r_y \omega) r_x \hat{\chi}_1 \hat{\chi}_2 (\bar{q} + r_x \hat{\chi}_2) - \lambda \hat{\varphi}^2 r_y^2 \hat{\chi}_2 (\bar{q} + r_x \hat{\chi}_2) -$$

$$-(\hat{q} - i\hat{\varphi} r_y \omega)\hat{\chi}_1 (\bar{q} + r_x \hat{\chi}_2)^2 + \kappa^2 \hat{\varphi}^2 r_y^2 (\hat{q} - i\hat{\varphi} r_y \omega)(\hat{\chi}_2 - \hat{\chi}_1) = 0. \tag{4.17}$$

We remind that $\hat{\chi}_2$ is found from

$$M^2(\bar{q}+r_x\hat{\chi}_2)^2+\hat{\varphi}^2r_y^2=r_x^2\hat{\chi}_2^2.$$

all solutions to which are described by

$$M(\bar{q}+r_x\hat{\chi}_2)=\hat{\varphi}r_y\frac{1-z^2}{2z},$$

$$r_x\hat{\chi}_2=\hat{\varphi}r_y\frac{1+z^2}{2z}.$$

where z is a number.

It is easy to show that equation (4.17) has the root $z^2=1$. Put $z=-\text{sign}\hat{\varphi}$. Then

$$\bar{q}=|\hat{\varphi}|r_y,$$

$$\hat{\chi}_0=\hat{\chi}_1=\hat{\chi}_2=-\frac{r_y|\hat{\varphi}|}{r_x},$$

$$\text{Re}\bar{q}=\text{Re}\hat{q}>0,\quad \text{Re}\hat{\chi}_{0,1,2}<0$$

and

$$|\chi_0|^2=1+2\varepsilon\text{Re}\hat{\chi}_0+\ldots<1,$$

$$|\chi_1|^2=1+2\varepsilon\text{Re}\hat{\chi}_1+\ldots<1,$$

$$|\chi_2|^2=1+2\varepsilon\text{Re}\hat{\chi}_2+\ldots<1,$$

$$|q|^2=1+2\varepsilon\text{Re}\hat{q}+\ldots>1.$$

The last inequalities mean that the difference Hadamard examples exist.

Remark 4.1. The considered example is aimed to illustrate importance of appropriate approximations of boundary conditions while applying difference schemes based on the splitting.

Results from the paper [14] are used in this section.

5. Construction of adequate difference scheme for problem II

We again consider **mixed problem II** without the equation on S and pose the following model for it:

$$\{A\cdot\tilde{T}+B\cdot\tilde{\xi}+C_0\cdot\tilde{\eta}\}\mathbb{U}=0,k,i,|j|=0,1,\ldots, \tag{5.1}$$

$$u_{0j}^k+d\cdot p_{0j}^k=0,\quad \tilde{T}v_{0j}^k-\lambda\tilde{\eta}p_{0j}^k=0,\quad k,|j|=0,1,\ldots, \tag{5.2}$$

$$\mathbb{U}_{ij}^0=\mathbb{U}_0(i\cdot h_x,j\cdot h_y),i,|j|=0,1,\ldots, \tag{5.3}$$

Here $\tilde{\mathbb{T}}=\tilde{\tau}+\omega\tilde{\eta}$, $\tilde{\xi}=L_t\dfrac{\xi}{h_x}$, $L_t=\alpha\varphi+\delta$ $(\alpha,\delta\geq 0,\alpha+\delta=1)$, $\tilde{\eta}=\dfrac{\bar{\eta}}{h_y}$, $\tilde{\tau}=\dfrac{\tau}{\Delta}$

The following two facts are valid both for **mixed problem II** ([4]; see also **introduction**) and for difference model (5.1)–(5.3):

a) multiply (5.1) scalarly by the vector

$$\begin{pmatrix} M^2(\tilde{T}+\tilde{\xi}) \\ -\tilde{\xi} \\ -\tilde{\eta} \end{pmatrix},$$

its elements are difference operators. After simple calculations we obtain that the component p of the vector $\mathbf{U}$ satisfies the difference wave equation:

$$\{M^2(\tilde{T}+\tilde{\xi}) - \tilde{\xi}^2 - \tilde{\eta}^2\}p = 0; \tag{5.4}$$

b) multiply (5.1) scalarly by

$$\begin{pmatrix} M^2\tilde{T} \\ -\tilde{T} \\ 0 \end{pmatrix}.$$

Consider the obtained expression at $i = 0$. With account of relations

$$\tilde{T}^2 u_{0j}^k = -d\tilde{T}^2 p_{0j}^k,$$

$$\tilde{T}\tilde{\eta} v_{0j}^k = \lambda\tilde{\eta}^2 p_{0j}^k,$$

which are the consequence of (5.2), we arrive at

$$\{M^2(1+d)\tilde{T}^2 - \beta^2\tilde{T}\tilde{\xi} + M^2\lambda\tilde{\eta}^2\}p_{0j}^k = 0. \tag{5.5}$$

If p satisfies (5.4) then

$$\mathbf{Y} = \begin{pmatrix} \tilde{L}_1 \\ L_2 \\ \tilde{L}_3 \end{pmatrix} p$$

satisfies the system (**see introduction**):

$$\{D\cdot\tilde{T} - Q\cdot\tilde{\xi} - R\cdot\tilde{\eta}\}\mathbf{Y} = 0, \tag{5.6}$$

Here

$$\tilde{L}_1 = M\cdot L_1, \quad L_1 = \frac{1}{\beta^2}\tilde{T}, \quad L_2 = \tilde{\xi} - \frac{M^2}{\beta^2}\tilde{T},$$

$$\tilde{L}_3 = \frac{L_3}{\beta^2}, \quad L_3 = \tilde{\eta}, \quad D = \frac{1}{\beta^2}(E + M^2 Q),$$

$$E = M\begin{pmatrix} 1 & m_1 & 0 \\ m_1 & 1 & 0 \\ 0 & 0 & 1 \end{pmatrix}, \quad Q = \begin{pmatrix} m_1 & 1 & 0 \\ 1 & m_1 & 0 \\ 0 & 0 & -m_1 \end{pmatrix},$$

$$R = \begin{pmatrix} 0 & 0 & 1 \\ 0 & 0 & m_1 \\ 1 & m_1 & 0 \end{pmatrix},$$

m_1 is a constant. We note that $D > 0$ if $|m_1| < 1$.

Compose an extended system from two systems of the form (5.6).

$$\left\{\begin{pmatrix} D & 0_3 \\ 0_3 & D \end{pmatrix}\tilde{T} - \begin{pmatrix} Q & O_3 \\ O_3 & Q \end{pmatrix}\tilde{\xi} - \begin{pmatrix} R & O_3 \\ O_3 & R \end{pmatrix}\tilde{\eta}\right\}\mathbf{X} = 0, \tag{5.7}$$

where

$$\mathbf{X} = \{\tilde{L}_1 L_+, L_2 L_+, \tilde{L}_3 L_+, \tilde{L}_1 L_-, L_2 L_-, \tilde{L}_3 L_-\}^*,$$

$$L_\pm = (L_\pm)^k_{ij} = (a_1 L_1 + a_2^\pm L_2) p^k_{ij},$$

$$a_1 = m\kappa, a_2^\pm = -\frac{n}{a_\pm}, a_\pm = \frac{M(\gamma \pm \sqrt{\gamma^2 - 4mn})}{2m},$$

$$m = \beta d + \frac{\lambda M^2}{\beta}, \quad n = -\frac{\lambda}{\beta}, \quad \gamma = \beta \cdot \kappa^2, \quad \gamma^2 > 4mn.$$

Boundary condition (5.5) can be rewritten as follows:

$$\{\tilde{L}_1 - a_\pm L_2\}(L_\pm)^k_{0j} = 0, k, |j| = 0, 1, \ldots. \tag{5.5'}$$

Let $m_1 = \dfrac{a_+}{1 + a_+^2}$ for the first system of the type (5.6) and $m_1 = \dfrac{a_-}{1 + a_-^2}$ for the second one, both these systems form the extended system (5.7). Then boundary conditions (5.5′) are dissipative for (5.7) because

$$\left(\begin{bmatrix} Q & 0_3 \\ 0_3 & Q \end{bmatrix}\mathbf{X_0}, \mathbf{X_0}\right) =$$

$$= a_+(a_1 L_1 L_2 p_0 + a_2^+ L_2^2 p_0)^2 + \frac{a_+(a_1 L_1 \tilde{L}_3 p_0 + a_2^+ L_2 \tilde{L}_3 p_0)^2}{(1 + a_+^2)} +$$

$$a_-(a_1 L_1 L_2 p_0 + a_2^- L_2^2 p_0)^2 + \frac{a_-(a_1 L_1 \tilde{L}_3 p_0 + a_2^- L_2 \tilde{L}_3 p_0)^2}{(1 + a_-^2)} > 0 \tag{5.8}$$

Present (5.1) in the form:

$$(A + \alpha r_x B\xi)\hat{\mathbb{U}} = (A - \delta r_x B\xi - C r_y \bar{\eta})\mathbb{U}. \tag{5.1'}$$

Let $\alpha = 1$, $\delta = 0$. Then it follows from (5.1′) that

$$([I_3 + r_x A^{-1} B\xi]\hat{\mathbb{U}}, [A + r_x B\xi]\hat{\mathbb{U}} = ([I_3 - r_y A^{-1} C\bar{\eta}]\mathbb{U}, [A - r_y C\bar{\eta}]\mathbb{U}).$$

Hence

$$(\hat{\mathbb{U}}, A\hat{\mathbb{U}}) + r_x\xi(\hat{\mathbb{U}}, B\hat{\mathbb{U}}) + r_x(\xi\hat{\mathbb{U}}, [r_x B A^{-1} B - B]\xi\hat{\mathbb{U}}) +$$
$$+ r_y(\bar{\eta}\mathbb{U}, [C - r_y C A^{-1} C]\bar{\eta}\mathbb{U}) + r_y\bar{\eta}(\mathbb{U}, C\mathbb{U}) - (\mathbb{U}, A\mathbb{U}) = 0. \tag{5.9}$$

If

$$\omega \geq \kappa, \quad r_x \geq 1, \quad r_y \geq \frac{M}{(\omega M + 1)}, \tag{5.10}$$

then from (5.9) we derive

$$||\hat{\mathbb{U}}||_A^2 - ||\mathbb{U}||_A^2 - \Delta h_y \sum_{j=-\infty}^{\infty} (\hat{\mathbb{U}}_0, B\hat{\mathbb{U}}_0) \leq 0, \tag{5.11}$$

where

$$||\mathbb{U}||_A^2 = h_x h_y \sum_{i=0}^{\infty} \sum_{j=-\infty}^{\infty} (\mathbb{U}, A\mathbb{U}),$$

and so on.

$$(\hat{\mathbb{U}}_0, B\hat{\mathbb{U}}_0) = (1 - 2Md + M^2d^2)(\hat{p}_0)^2 + M^2(\hat{v}_0)^2$$

(by virtue of (5.2)).

The third equation from (5.1) after application of the difference operator $\tilde{T}$ gives

$$(1 + r_x\xi)\tilde{T}\hat{v} = \tilde{T}v - \omega r_y \bar{\eta}\tilde{T}v - \Delta\tilde{T}\tilde{\eta}p\kappa^2.$$

Mulitiply this equality by $(1 + r_x\xi)\tilde{T}\hat{v}$. With account of (5.10), we obtain the estimation:

$$||\tilde{T}\hat{v}||^2 - ||\tilde{T}v||^2 - \Delta h_y \sum_{j=-\infty}^{\infty} (\tilde{T}\hat{v})^2 \leq \Delta\mu_1||\tilde{T}v||^2 + ||\tilde{T}\tilde{\eta}p||^2, \tag{5.12}$$

where

$$||\tilde{T}v||^2 = h_x h_y \sum_{j=0}^{\infty} \sum_{j=-\infty}^{\infty} (\tilde{T}v)^2$$

and so on,

$$\tilde{T}\hat{v}_0 = \lambda\tilde{\eta}\hat{p}_0$$

(by virtue of (5.2)), $\mu_1 > 0$ is a constant.

Turn back to problem (5.7), (5.5′). Instead of system (5.7) we will use its corollary

$$\left\{ \begin{pmatrix} D_1 & 0_4 \\ 0_4 & D_1 \end{pmatrix} \tilde{T} - \begin{pmatrix} Q_1 & 0_4 \\ 0_4 & Q_1 \end{pmatrix} \tilde{\xi} - \begin{pmatrix} R_1 & 0_4 \\ 0_4 & R_1 \end{pmatrix} \tilde{\eta} \right\} \mathbb{W} = 0. \tag{5.7′}$$

Here

$$D_1 = \frac{M}{\beta^2} \cdot \begin{pmatrix} 1 & -M \\ -M & 1 \end{pmatrix} \otimes H, \quad Q_1 = \begin{pmatrix} 0 & -1 \\ -1 & 0 \end{pmatrix} \otimes H,$$

$$R_1 = \begin{pmatrix} -1 & 0 \\ 0 & 1 \end{pmatrix} \otimes H, H = \begin{pmatrix} 1 & -m_1 \\ -m_1 & 1 \end{pmatrix}$$

$$(H > 0, \text{ if } |m_1| < 1),$$

$$\mathbb{W} = \mathbb{W}_{ij}^k = \begin{pmatrix} T_0 & 0 \\ 0 & T_0 \end{pmatrix} \cdot \mathbf{X},$$

$$T_0 = \frac{1}{\sqrt{2}} \begin{pmatrix} 1 & 0 & -1 \\ 0 & -1 & 0 \\ 0 & -1 & 0 \\ 1 & 0 & 1 \end{pmatrix}.$$

We note that $A \otimes B$ is the Kronecker product of the matrices A, B (see [15]). Since the matrix coefficients in (5.7′) are block-diagonal matrices we will consider systems of the type

$$\{D_1\tilde{T} - Q_1\tilde{\xi} + R_2\tilde{\eta}\}\mathbb{A} = 0, \tag{5.13}$$

where $\mathbb{A}$ is a vector of the corresponding components of $\mathbb{W}$, $R_2 = \omega D_1 - R_1$. Rearranging system (5.13) into

$$(D_1 - r_x Q_1 \xi)\hat{\mathbb{A}} = (D_1 - r_y R_2 \bar{\eta})\mathbb{A},$$

we arrive, as before, at the estimation:

$$\|\hat{\mathbb{W}}\|^2_{D_1} \leq \|\mathbb{W}\|^2_{D_1}, \tag{5.14}$$

where

$$\|\mathbb{W}\|^2_{D_1} = h_x \cdot h_y \sum_{i=0}^{\infty} \sum_{j=-\infty}^{\infty} \left(\begin{bmatrix} D_1 & 0_4 \\ 0_4 & D_1 \end{bmatrix} \cdot \mathbb{W}, \mathbb{W} \right) =$$

$$= h_x \cdot h_y \sum_{i=0}^{\infty} \sum_{j=-\infty}^{\infty} \left(\begin{bmatrix} D & 0_3 \\ 0_3 & D \end{bmatrix} \cdot \mathbf{X}, \mathbf{X} \right) = \|\mathbf{X}\|^2_D.$$

While deriving (5.14) we used (5.8), under the assumption that (5.10) is valid.

Since

$$L_1 L_2 p = \frac{L_1 L_+ - L_1 L_-}{a_2^+ - a_2^-}, \quad L_1^2 p = \frac{L_1 L_+ - a_2^+ L_1 L_2 p}{a_1},$$

$$L_2^2 p = \frac{L_2 L_+ - L_2 L_-}{a_2^+ - a_2^-}, \quad \tilde{L}_3^2 p = M^2 L_1^2 p - L_2^2 p,$$

$$L_2 \tilde{L}_3 p = \frac{\tilde{L}_3 L_+ - \tilde{L}_3 L_-}{a_2^+ - a_2^-}, \quad L_1 \tilde{L}_3 p = \frac{\tilde{L}_3 L_+ - a_2^+ L_2 \tilde{L}_3 p}{a_1},$$

the aggregates

$$\tilde{T}^2 p, \quad \tilde{T}\tilde{\xi}p, \quad \tilde{T}\tilde{\eta}p, \quad \tilde{\xi}^2 p, \quad \tilde{\xi}\tilde{\eta}p, \quad \tilde{\eta}^2 p$$

can be estimated with the help of (5.14).

We finally write down the evident relations:

$$\tilde{\tau}p - \tilde{\xi}p + \omega\tilde{\eta}p - \tilde{T}p + \tilde{\xi}p = 0,$$

$$\tilde{\tau}(\tilde{T}p) - \tilde{\xi}(\tilde{T}p) + \omega\tilde{\eta}(\tilde{T}p) - \tilde{T}^2 p + \tilde{T}\tilde{\xi}p = 0,$$

$$\tilde{\tau}(\tilde{\xi}p) - \tilde{\xi}(\tilde{\xi}p) + \omega\tilde{\eta}(\tilde{\xi}p) - \tilde{T}\tilde{\xi}p + \tilde{\xi}^2 p = 0, \tag{5.15}$$

$$\tilde{\tau}(\tilde{\eta}p) - \tilde{\xi}(\tilde{\eta}p) + \omega\tilde{\eta}(\tilde{\eta}p) - \tilde{T}\tilde{\eta}p + \tilde{\xi}\tilde{\eta}p = 0,$$
$$\tilde{\tau}v - \tilde{\xi}v + \omega\tilde{\eta}v - 2\tilde{T}v - \kappa^2\tilde{\eta}p = 0.$$

Introduce the aggregate

$$\tilde{J}_k = \varepsilon_1\{||\mathbb{U}||_A^2 + ||\tilde{T}v||^2\} + ||\mathbf{X}||_D^2 + ||p||^2 + ||\tilde{T}p||^2 + ||\tilde{\xi}p||^2 + ||\tilde{\eta}p||^2 + ||v||^2,$$

where $\varepsilon_1 > 0$ is a constant. From (5.11),(5.12), (5.14), (5.15) it follows that

$$\tilde{J}_{k+1} - \tilde{J}_k \leq k_1 \cdot \Delta \cdot J_k,$$

where

$$J_k = ||\mathbb{U}||^2 + ||\tilde{T}v||^2 + ||\tilde{T}p||^2 + ||\tilde{\xi}p||^2 + ||\tilde{\eta}p||^2 + ||\tilde{T}^2p||^2 +$$
$$+||\tilde{T}\tilde{\xi}p||^2 + ||\tilde{T}\tilde{\eta}p||^2 + ||\tilde{\xi}^2p||^2 + ||\tilde{\xi}\tilde{\eta}p||^2 + ||\tilde{\eta}^2p||^2,$$
$$||\mathbb{U}||^2 = h_x \cdot h_y \sum_{i=0}^{\infty} \sum_{j=-\infty}^{\infty} (\mathbb{U}, \mathbb{U}).$$

$k_1 > 0$ is a constant.

Thus, we obviously derive the desired estimation (the difference analog of estimation (0.15) from **introduction**

$$J_k \leq \text{const} \cdot J_0, k = 1.2, \ldots,$$

and establish stability of the difference scheme. The sufficient conditions are presented by (5.10).

Remark 5.1 The difference model of **problem II** from this section is inconvenient to be used for practical calculations because of very tough restrictions on steps of the difference net (see (5.10)). However, the one-dimensional variant of this scheme was successfully used by the authors for calculation of mixed problem for quasilinear gas dynamics equations [16].

6. Investigation of some differential-difference models for problem II

Results from sections 3–5 confirm that construction of stable difference models for **mixed problem II** is a complicated problem. The main point is approximation of boundary conditions such that the difference model in whole stays stable. In this connection the idea to avoid quantization with respect to x (at least theoretically) seems to be interesting. As a result we come to the so-called difference-differential schemes.

In the domain R^3_{++} we carry out the **quantization** over t and y with the steps $\Delta t = \Delta$, $\Delta y = h_y$. We introduce the notation (see also notation from section 1):

$$\mathbb{U}(k \cdot \Delta, x, j \cdot h_y) = \mathbb{U}_j^k(x) = \mathbb{U}, \quad \varphi\mathbb{U}_j^k(x) = \varphi\mathbb{U} = \mathbb{U}_j^{k+1}(x),$$

$$\theta\mathbb{U}_j^k(x) = \theta\mathbb{U} = \mathbb{U}_{j+1}^k(x), \quad \theta^{-1}\mathbb{U}_j^k(x) = \theta^{-1}\mathbb{U} = \mathbb{U}_{j-1}^k(x),$$

$$L_t \mathbb{U}_j^k(x) = L_t \mathbb{U} = \tilde{\mathbb{U}}, \quad L_t = \alpha\varphi + \delta, \quad \hat{\tau} = \frac{\varphi - 1}{\Delta}, \quad \xi = \frac{\partial}{\partial x},$$

$$\hat{\xi} = L_t \xi, \eta = \theta - 1, \quad \bar{\eta} = 1 - \theta^{-1}, \quad \hat{\eta}_0 = \frac{\eta + \bar{\eta}}{2h_y}, \hat{\eta} = L_t \hat{\eta}_0,$$

where α, $\delta \geq 0$ ($\alpha + \delta = 1$ are constants).

We formulate now the difference-differential model to the initial mixed problem

$$A\hat{\tau}\mathbb{U} + B\hat{\xi}\mathbb{U} + C \cdot \hat{\eta}\mathbb{U} = 0, \quad x > 0, k, |j| = 0, 1, \ldots; \tag{6.1}$$

$$\begin{cases} u_j^k(0) + dp_j^k(0) = 0, \\ \hat{\tau} v_j^k + \omega\hat{\eta} v_j^k(0) - \lambda\hat{\eta} p_j^k(0) = 0, \end{cases} \tag{6.2}$$

$$k, |j| = 0, 1, \ldots;$$

$$\mathbb{U}_j^0(x) = \mathbb{U}_0(x, j \cdot h_y), \quad x \geq 0, \quad |j| = 0, 1, \ldots. \tag{6.3}$$

Carried out quantization over t in R^3_{++}, we have one more difference-differential model

$$A\hat{\tau}\mathbb{U}^k + B\hat{\xi}\mathbb{U}^k + C \cdot \hat{\eta}\mathbb{U}^k = 0, \quad (x, y) \in R^2_+, \quad k = 0, 1, \ldots; \tag{6.4}$$

$$\begin{cases} u^k + dp^k = 0, \\ \hat{\tau} v^k + \omega\hat{\eta} v^k - \lambda\hat{\eta} p^k = 0, \end{cases} \tag{6.5}$$

$$x = 0, \quad y \in R^1, \quad k = 0, 1, \ldots;$$

$$\mathbb{U}^0(x, y) = \mathbb{U}_0(x, y), \quad x \geq 0, \quad y \in R^1. \tag{6.6}$$

Here

$$\mathbb{U}(k \cdot \Delta, x, jh_y) = \mathbb{U}^k(x, y) = \mathbb{U}^k, \quad \hat{\eta} = L_t \eta, \quad \eta = \frac{\partial}{\partial y}.$$

We will call **difference-differential model** (6.1)–(6.3) by **model I**, and **difference-differential model** (6.4)–(6.6) by **model II**.

The following facts are valid for the difference-differential **models I and II** (see [4], **introduction** and section 5). Multiply (6.1) (or (6.4)) scalarly by

$$\begin{pmatrix} M^2(\hat{T} + \hat{\xi}) \\ -\hat{\xi} \\ -\hat{\eta} \end{pmatrix},$$

where $\hat{T} = \hat{\tau} + \omega\hat{\eta}$. After some transformations we obtain that the component p of the vector $\mathbb{U}$ satisfies the difference-differential wave equation:

$$\{M^2(\hat{T} + \hat{\xi}) - \hat{\xi}^2 - \hat{\eta}^2\} p = 0. \tag{6.7}$$

Multiply (6.1) scalarly by

$$\begin{pmatrix} M^2\hat{T} \\ -\hat{T} \\ 0 \end{pmatrix}.$$

Put $x = 0$ in this expression:

$$\{M^2(1+d)\hat{T}^2 - \beta^2\hat{T}\hat{\xi} + M^2\lambda\hat{\eta}^2\}p_j^k(0) = 0. \tag{6.8}$$

Then instead of problem (6.1)–(6.3) we come to another difference-differential model:

$$\begin{cases} \{M^2(\hat{T}+\hat{\xi})^2 - \hat{\xi}^2 - \hat{\eta}^2\}p = 0, \\ M^2(\hat{T}+\hat{\xi})u = -\hat{\xi}p, \\ M^2(\hat{T}+\hat{\xi})v = -\hat{\eta}p, x > 0, k, |j| = 0, 1, \ldots; \end{cases} \tag{6.9}$$

$$\begin{cases} \{M^2(1+d)\hat{T}^2 - \beta^2\hat{T}\hat{\xi} + M^2\lambda\hat{\eta}^2\}p_j^k(0) = 0, \\ u_j^k(0) = -dp_j^k(0), \\ \hat{T}v_j^k(0) = \lambda\hat{\eta}p_j^k(0) = 0, k, |j| = 0, 1, \ldots; \end{cases} \tag{6.10}$$

$$\begin{cases} \mathrm{U}_j^0(x) = \mathrm{U}_0(x, j \cdot h_y), \\ (\hat{T}+\hat{\xi})p_j^0(x) + \hat{\xi}u_j^0(x) + \hat{\eta}v_j^0(x) = 0, \end{cases} \tag{6.11}$$

$$x \geq 0, |j| = 0, 1, \ldots.$$

We will show that under certain conditions problems (6.1)–(6.3) and (6.9)–(6.11) are equivalent. Indead, it follows from (6.9) that

$$\{M^2(\hat{T}+\hat{\xi})^2 - \hat{\xi}^2 - \hat{\eta}^2\}p = M^2(\hat{T}+\hat{\xi})\mathcal{P} = 0,$$

where

$$\mathcal{P} = \mathcal{P}_j^k(x) = (\hat{T}+\hat{\xi})p_j^k(x) + \hat{\xi}u_j^k(x) + \hat{\eta}v_j^k(x).$$

By analogy, from из (6.10)

$$\{M^2(1+d)\hat{T}^2 - \beta^2\hat{T}\hat{\xi} + M^2\lambda\hat{\eta}^2\}p_j^k(0) = M^2\hat{T}\mathcal{P}_j^k(0) = 0.$$

Thus, we formulate a problem for the aggregate $\mathcal{P}$.

$$(\hat{T}+\hat{\xi})\mathcal{P}_j^k(x) = 0, \quad x > 0, \quad k, |j| = 0, 1, \ldots; \tag{6.12}$$

$$\hat{T}\mathcal{P}_j^k(0) = 0, \quad k, |j| = 0, 1, \ldots; \tag{6.13}$$

$$\mathcal{P}_j^0(x) = 0, \quad x \geq 0, \quad |j| = 0, 1, \ldots. \tag{6.14}$$

It follows from (6.13) that

$$\hat{T}\mathcal{P}_j^k(0) = \hat{\tau}\mathcal{P}_j^k(0) + \omega\hat{\eta}_0\tilde{\mathcal{P}}_j^k(0) = 0.$$

Multiply this relation by $2\hat{\mathcal{P}}_j^k(0)$ and take into consideration the evident relation

$$2\mathcal{P}_j^k(0)\hat{\tau}\mathcal{P}_j^k(0) = \hat{\tau}(\mathcal{P}_j^k(0))^2 - \Delta(\hat{\tau}\mathcal{P}_j^k(0))^2,$$

$$2\hat{\mathcal{P}}_j^k(0)\hat{\eta}_0\hat{\mathcal{P}}_j^k(0) = \hat{\eta}_0(\hat{\mathcal{P}}_j^k(0))^2 - \frac{\eta}{2h_y}(\bar{\eta}\hat{\mathcal{P}}_j^k(0))^2,$$

we finally obtain

$$\hat{\tau}(\mathcal{P}_j^k(0))^2 + \omega\hat{\eta}_0(\tilde{\mathcal{P}}_j^k(0))^2 + (\alpha-\delta)\Delta(\hat{\tau}\mathcal{P}_j^k(0))^2 - \omega\frac{\eta}{2h_y}(\bar{\eta}\hat{\mathcal{P}}_j^k(0))^2 = 0.$$

Thus at $\alpha \geq \delta$ the inequality

$$||\mathcal{P}^{k+1}||_0^2 \leq ||\mathcal{P}^k||_0^2,$$

is true where $||\mathcal{P}^k||_0^2 = h_y \sum_{j=-\infty}^{\infty}(\mathcal{P}_j^k(0))^2$. By (6.14), $||\mathcal{P}^0||_0^2 = 0$, then $\mathcal{P}_j^k(x_0) = 0$, $k, |j| = 0, 1, \ldots$. By analogy, presenting (6.12) in the form

$$\hat{\tau}\mathcal{P} + \xi\tilde{\mathcal{P}} + \omega\hat{\eta}_0\tilde{\mathcal{P}} = 0,$$

and multiplying it by $2\tilde{\mathcal{P}}$, we come to the expression

$$\hat{\tau}(\mathcal{P})^2 + \xi(\tilde{\mathcal{P}})^2 + \omega\hat{\eta}_0(\hat{\mathcal{P}})^2 + (\alpha-\delta)\Delta(\hat{\tau}\mathcal{P})^2 - \omega\frac{\eta}{2h_y}(\bar{\eta}\tilde{\mathcal{P}})^2 = 0.$$

Hence, under the assumption $\alpha \geq \delta$, we obtain

$$||\mathcal{P}^{k+1}||^2 \leq ||\mathcal{P}^k||^2,$$

where $||\mathcal{P}^k||^2 = h_y \sum_{j=-\infty}^{\infty} \int_0^{\infty}(\mathcal{P}_j^k(x))^2dx$. By (6.14), $||\mathcal{P}^0||_0^2 = 0$, then $\mathcal{P}_j^k(x) \equiv 0$, $x \geq 0$, $k, |j| = 0, 1, \ldots$. Thus, the first equation from (6.1) holds everywhere. Consequently, problems (6.1)–(6.3) and (6.9)–(6.11) are equivalent.

If p satisfies (6.7) then the vector

$$\mathbf{Y}_j^k(x) = Y = \begin{bmatrix} \tilde{L}_1 \\ L_2 \\ \tilde{L}_3 \end{bmatrix} p$$

satisfies the system ([4], see also section 5)

$$\{D \cdot \hat{T} - Q \cdot \hat{\xi} - R \cdot \hat{\eta}\}\mathbf{Y} = 0, \tag{6.15}$$

Here

$$L_1 = \frac{1}{\beta^2}\hat{T}, \quad L_2 = \hat{\xi} - \frac{M^2}{\beta^2}\hat{T}, \quad L_3 = \hat{\eta},$$

$$\tilde{L}_1 = ML_1, \quad \tilde{L}_3 = \frac{1}{\beta}L_3.$$

$$E = M\begin{pmatrix} 1 & m_1 & 0 \\ m_1 & 1 & 0 \\ 0 & 0 & 1 \end{pmatrix}, \quad Q = \begin{pmatrix} m_1 & 1 & 0 \\ 1 & m_1 & 0 \\ 0 & 0 & -m_1 \end{pmatrix},$$

$$R = \begin{pmatrix} 0 & 0 & 1 \\ 0 & 0 & m_1 \\ 1 & m_1 & 0 \end{pmatrix}, \quad D = \frac{1}{\beta^2}(E + M^2Q),$$

m_1 is a constant. We note that $D > 0$ if $|m_1| < 1$.

Compose an extended system from the remained two systems of the form (6.15):

$$\left\{\begin{pmatrix} D & O_3 \\ O_3 & D \end{pmatrix}\hat{T} - \begin{pmatrix} Q & O_3 \\ O_3 & Q \end{pmatrix}\tilde{\xi} - \begin{pmatrix} R & O_3 \\ O_3 & R \end{pmatrix}\hat{\tilde{\eta}}\right\}\mathbf{X} = 0, \tag{6.16}$$

where

$$\mathbf{X}_j^k(x) = \mathbf{X} = \{\tilde{L}_1 L_+, L_2 L_+, \tilde{L}_3 L_+, \tilde{L}_1 L_-, L_2 L_-, \tilde{L}_3 L_-\}^*,$$

$$L_\pm = (L_\pm)_j^k(x) = (a_1 L_1 + a_2^\pm L_2)p_j^k(x),$$

$$a_1 = m\kappa, a_2^\pm = -\frac{n}{a_\pm}, \quad \gamma = \beta\kappa^2, \quad n = -\frac{\lambda}{\beta},$$

$$a_\pm = \frac{M(\gamma \pm \sqrt{\gamma^2 - 4mn})}{2m}, \quad m = \beta d + \frac{\lambda M^2}{\beta}, \quad \kappa = M^{-1}$$

Condition (6.8) can be presented in the form:

$$\{\tilde{L}_1 - a_\pm L_2\}/(L_\pm)_j^k(0) = 0, \quad k, |j| = 0, 1, \ldots. \tag{6.8$'$}$$

Let $m_1 = \frac{a_+}{1 + a_+^2}$ for the first system and $m_1 = \frac{a_-}{1+a_-^2}$ for the second system of the form (6.15) which form the extended system (6.16). Then boundary conditions (6.8$'$) are **dissipative** for system (6.16) since

$$\left(\begin{bmatrix} Q & O_3 \\ O_3 & Q \end{bmatrix}\mathbf{X}_j^k(0), \mathbf{X}_j^k(0)\right) = a_+(a_1 L_1 L_2 p_j^k(0) + a_2^+ L_2^2 p_j^k(0))^2 +$$

$$+\frac{a_+}{1 + a_+^2}(a_1 L_1 \tilde{L}_3 p_j^k(0) + a_2^+ L_2 \tilde{L}_3 p_j^k(0))^2 + a_-(a_1 L_1 L_2 p_j^k(0) + a_2^- L_2^2 p_j^k(0))^2 +$$

$$+\frac{a_-}{1 + a_-^2}(a_1 L_1 \tilde{L}_3 p_j^k(0) + a_2^- L_2 \tilde{L}_3 p_j^k(0))^2 > 0. \tag{6.17}$$

Remark 6.1. The results from above hold valid for the difference-differential **model II**.

It is convenient to use instead of (6.1) its corollary:

$$M\hat{\tau}\mathbf{Z} + B_1\hat{\xi}\mathbf{Z} + C_1\hat{\eta}\mathbf{Z} + \mathbf{\Gamma} = 0, \tag{6.1$'$}$$

where $\mathbf{Z}_j^k(x) = Z = T_0 A^{\frac{1}{2}}\mathbf{U}$,

$$T_0 = \frac{1}{\sqrt{2}}\begin{bmatrix} 1 & 0 & -1 \\ 0 & -1 & 0 \\ 0 & -1 & 0 \\ 1 & 0 & 1 \end{bmatrix}, \quad B_1 = \begin{bmatrix} M & -1 \\ -1 & M \end{bmatrix} \times I_2,$$

$$C_1 = \begin{bmatrix} \omega M - 1 & 0 \\ 0 & \omega M + 1 \end{bmatrix} \times I_2, \quad I_2 = \begin{bmatrix} 1 & 0 \\ 0 & 1 \end{bmatrix},$$

$$\Gamma_j^k(x) = \Gamma = \frac{M}{\sqrt{2}} \begin{bmatrix} 0 \\ -1 \\ 1 \\ 0 \end{bmatrix} \Omega, \quad \Omega_j^k(x) = \Omega = \hat{\eta} u - \hat{\xi} v.$$

We note that, by (6.1), the aggregate Ω (the difference-differential analog of the velocity vortex) satisfies the relation:

$$(\hat{T} + \hat{\xi})\Omega = 0, \quad x > 0, \quad k, |j| = 0, 1, \ldots. \tag{6.18}$$

Multiply (6.1′) scalarly by $2\tilde{\mathbf{Z}}$ and take into account the relations

$$(2\tilde{\mathbf{Z}}, A\hat{\tau}\mathbf{Z}) = \hat{\tau}(\mathbf{Z}, A\mathbf{Z}) + (\alpha - \delta)\Delta(\hat{\tau}\mathbf{Z}, A\hat{\tau}\mathbf{Z}),$$

$$(2\tilde{\mathbf{Z}}, C\hat{\tau}\tilde{\mathbf{Z}}) = \hat{\eta}_0(\tilde{\mathbf{Z}}, C\tilde{\mathbf{Z}}) + \frac{\eta}{2h_y}(\bar{\eta}\tilde{\mathbf{Z}}, C\bar{\eta}\tilde{\mathbf{Z}}),$$

$$(2\tilde{\mathbf{Z}}, B\hat{\xi}\tilde{\mathbf{Z}}) = \hat{\xi}(\tilde{\mathbf{Z}}, B\tilde{\mathbf{Z}}),$$

(A,B,C are symmetric matrices!), as the result we obtain

$$M\hat{\tau}(|\mathbf{Z}|^2) + \hat{\xi}(\tilde{\mathbf{Z}}, B_1\tilde{\mathbf{Z}}) + \hat{\eta}_0(\tilde{\mathbf{Z}}, C_1\tilde{\mathbf{Z}}) + M(\alpha - \delta)\Delta(|\hat{\tau}\mathbf{Z}|^2) -$$

$$-\frac{\eta}{2h_y}(\bar{\eta}\tilde{\mathbf{Z}}, C_1\bar{\eta}\tilde{\mathbf{Z}}) + 2(\tilde{\mathbf{Z}}, \mathbf{\Gamma}) = 0, |\mathbf{Z}|^2 = (\mathbf{Z}, \mathbf{Z}).$$

By analogy, we multiply (6.18) by $2\tilde{\mathbf{\Omega}}$ and, thus, we have

$$\hat{\tau}(\mathbf{\Omega})^2 + \hat{\xi}(\tilde{\mathbf{\Omega}})^2 + \omega\hat{\eta}_0(\tilde{\mathbf{\Omega}})^2 + (\alpha - \delta)\Delta(\hat{\tau}\mathbf{\Omega})^2 - \omega\frac{\eta}{2h_y}(\bar{\eta}\tilde{\mathbf{\Omega}})^2 = 0.$$

In a view of these relations, we easily derive that if $\alpha \geq \delta$ then

$$\hat{\tau}\{||\mathbf{Z^k}||^2 + ||\mathbf{\Omega^k}||^2\} - h_y \sum_{j=-\infty}^{\infty} \{(\tilde{\mathbf{Z}}_j^k(0), B_1\tilde{\mathbf{Z}}_j^k(0)) + (\tilde{\mathbf{\Omega}}_j^k(0))^2\} \leq$$

$$\leq \tilde{N}_1\{||\mathbf{Z^k}||^2 + ||\mathbf{Z^{k+1}}||^2 + ||\mathbf{\Omega^k}||^2\}, \tag{6.19}$$

where

$$||\mathbf{Z^k}||^2 = Mh_y \sum_{j=-\infty}^{\infty} \int_0^{\infty} |\mathbf{Z}_j^k(x)|^2 dx,$$

$$||\mathbf{\Omega^k}||^2 = h_y \sum_{j=-\infty}^{\infty} \int_0^{\infty} (\mathbf{\Omega}_j^k(x))^2 dx,$$

While obtaining (6.19) we used the inequalities:

$$2|(\tilde{\mathbf{Z}}, \mathbf{\Gamma})| \leq 2|\tilde{\mathbf{Z}}||\mathbf{\Gamma}| \leq M|\tilde{\mathbf{Z}}|^2 + \kappa|\mathbf{\Gamma}|^2 \leq$$

$$\leq 2\alpha^2 M|\varphi\mathbf{Z}|^2 + 2\delta^2 M|\mathbf{Z}|^2 + M(\Omega)^2,$$

moreover $\tilde{N}_1 = \max\{2\alpha^2 M, M\}$. We also note that by virtue of (6.2)

$$(\tilde{\mathbf{Z}}_j^k(0), B_1\tilde{\mathbf{Z}}_j^k(0)) = M(1 + M^2d^2 - 2d)(\tilde{p}_j^k(0))^2 +$$

$$+M^3(\tilde{v}_j^k(0))^2, (\tilde{\Omega}_j^k(0))^2 = (\lambda + \kappa^2 - d)^2(\tilde{\eta}\tilde{p}_j^k(0))^2.$$

Remark 6.2. For difference-differential model II the estimation analogous to (6.19) ia the following:

$$\hat{\tau}\{||\mathbf{Z^k}||^2 + ||\boldsymbol{\Omega}^\mathbf{k}||^2\} - \int_R \{(\tilde{\mathbf{Z}}^k(0,y), B_1\tilde{\mathbf{Z}}^k(0,y)) + (\tilde{\boldsymbol{\Omega}}^k(0,y))^2\}dy \leq$$

$$\leq N_1\{||\mathbf{Z^k}||^2 + ||\mathbf{Z^{k+1}}||^2 + ||\boldsymbol{\Omega}^\mathbf{k}||^2\},$$

where

$$||\mathbf{Z^k}||^2 = M\int R_+^2 |\mathbf{Z}^k(x,y)|^2 dxdy,$$

$$||\boldsymbol{\Omega}^\mathbf{k}||^2 = \iint_{R_+^2} |\boldsymbol{\Omega}^k(x,y)|^2 dxdy.$$

It follows from the third equation of (6.1) that

$$(\hat{T} + \hat{\xi})\hat{T}v = -\hat{T}\hat{\eta}p\kappa^2.$$

Hence

$$\hat{\tau}(||\hat{T}v^k||^2) - h_y \sum_{j=-\infty}^{\infty} (\tilde{\hat{T}}v_j^k(0))^2 \leq \tilde{N}_2\{||\hat{T}\hat{\eta}p^k||^2 + ||\hat{T}v^k||^2 + ||\hat{T}v^{k+1}||^2\}, \quad (6.20)$$

where

$$||\hat{T}v^k||^2 = h_y \sum_{j=-\infty}^{\infty} \int_0^\infty |(\hat{T}v_j^k(x))^2 dx,$$

$$||\hat{T}\hat{\eta}p^k||^2 = h_y \sum_{j=-\infty}^{\infty} \int_0^\infty |(\hat{T}\hat{\eta}p_j^k(x))^2 dx,$$

$$\tilde{N}_2 = \max\{1, 2\alpha^2\}\kappa^2, \quad (\tilde{\hat{T}}v_j^k(0))^2 = \lambda^2((\tilde{\hat{\eta}}p_j^k(0))^2.$$

Remark 6.3. The analogous inequality for difference-differential model II is

$$\hat{\tau}(||\hat{T}v^k||^2) - \int_{R^1} (\tilde{\hat{T}}\tilde{v}^k(0,y))^2 dy \leq \tilde{N}_2\{||\hat{T}\hat{\eta}p^k||^2 + ||\hat{T}v^k||^2 + ||\hat{T}v^{k+1}||^2\}, \quad (6.20)$$

where

$$||\hat{T}v^k||^2 = \iint_{R_+^2} (\hat{T}v^k(x,y))^2 dxdy,$$

$$||\hat{T}\hat{\eta}p^k||^2 = \iint\limits_{R^2_+} (\hat{T}\hat{\eta}p^k(x,y))^2 dxdy.$$

We now come to consideration of problem (6.16), (6.8′). Instead of (6.16) we will use its corollary

$$\left\{\begin{pmatrix} D_1 & O_4 \\ O_4 & D_1 \end{pmatrix}\hat{T} - \begin{pmatrix} Q_1 & O_4 \\ O_4 & Q_1 \end{pmatrix}\hat{\xi} - \begin{pmatrix} R_1 & O_4 \\ O_4 & R_1 \end{pmatrix}\hat{\eta}\right\}\mathbf{W} = 0. \qquad (6.16')$$

Here

$$D_1 = \frac{M}{\beta^2}\cdot\begin{pmatrix} 1 & -M \\ -M & 1 \end{pmatrix}\otimes H, \quad Q_1 = \begin{pmatrix} 0 & -1 \\ -1 & 0 \end{pmatrix}\otimes H,$$

$$R_1 = \begin{pmatrix} -1 & 0 \\ 0 & 1 \end{pmatrix}\otimes H, \quad H = \begin{pmatrix} 1 & -m_1 \\ -m_1 & 1 \end{pmatrix} > 0, \quad m_1 < 1(!),$$

$$\mathbf{W} = \mathbf{W}^k_j(x) = \begin{pmatrix} T_0 & O \\ O & T_0 \end{pmatrix}\cdot\mathbf{X}.$$

The matrix coefficients in (6.16′) are block-diagonal, by this reason we will consider systems of the type:

$$\{D_1\hat{\tau} - Q_1\hat{\xi} + (\omega D_1 - R_1)\hat{\eta}\}\hat{\mathbf{W}} = 0,$$

where $\hat{\mathbf{W}}$ is the vector of the corresponding components of the vector $\mathbf{W}$. Multiplying the system scalarly by $2\tilde{\hat{\mathbf{W}}}$, we obtain

$$\hat{\tau}(\hat{\mathbf{W}}, D_1\hat{W}) - \hat{\xi}(\tilde{\hat{\mathbf{W}}}, Q_1\tilde{\hat{\mathbf{W}}}) + \hat{\eta}_0(\tilde{\hat{\mathbf{W}}}, (\omega D_1 - R_1)\tilde{\hat{\mathbf{W}}}) +$$

$$+(\alpha-\delta)\Delta(\hat{\tau}\hat{\mathbf{W}}, D_1\hat{\tau}\hat{\mathbf{W}}) - \frac{\eta}{2h_y}(\bar{\eta}\tilde{\hat{\mathbf{W}}}, (\omega D_1 - R_1)\bar{\eta}\tilde{\hat{\mathbf{W}}}) = 0,$$

At $\alpha > \delta$

$$\hat{\tau}||\mathbf{W}^k||^2_{D_1} + h_y\sum_{j=-\infty}^{\infty}(\tilde{\mathbf{W}}^k_j(0), \begin{pmatrix} Q_1 & O_4 \\ O_4 & Q_1 \end{pmatrix}\tilde{\mathbf{W}}^k_j(0)) \le 0,$$

where

$$||\mathbf{W}^k||^2_{D_1} = h_y\sum_{j=-\infty}^{\infty}\int_0^{\infty}\left(\begin{bmatrix} D_1 & O_4 \\ O_4 & D_1 \end{bmatrix}\cdot\mathbf{W}^k_j(x), \mathbf{W}^k_j(x)\right)dx =$$

$$= h_y\sum_{j=-\infty}^{\infty}\int_0^{\infty}\left(\begin{bmatrix} D & O_3 \\ O_3 & D \end{bmatrix}\cdot\mathbf{X}^k_j(x), \mathbf{X}^k_j(x)\right)dx = ||\mathbf{X}^k||^2_D.$$

Since, by virtue (6.17),

$$\left(\tilde{\mathbf{W}}^k_j(0), \begin{pmatrix} Q_1 & O_4 \\ O_4 & Q_1 \end{pmatrix}\tilde{\mathbf{W}}^k_j(0)\right) > 0,$$

finally we have the estimation

$$||\mathbf{X}^{k+1}||_D^2 \le ||\mathbf{X}^k||_D^2, \tag{6.21}$$

Remark 6.4. An analogous estimation can be inferred for **model II** also

$$||\mathbf{X}^{k+1}||_D^2 \le ||\mathbf{X}^k||_D^2,$$

where

$$||\mathbf{X}^k||_D^2 = \iint\limits_{R_+^2} \left(\begin{bmatrix} D & O_3 \\ O_3 & D \end{bmatrix} \cdot \mathbf{X}^k(x,y), \mathbf{X}^k(x,y) \right) dxdy.$$

We have

$$L_1 L_2 p = \frac{L_1 L_+ - L_1 L_-}{a_2^+ - a_2^-}, \quad L_1^2 p = \frac{L_1 L_+ - a_2^+ L_1 L_2 p}{a_1},$$

$$L_2^2 p = \frac{L_2 L_+ - L_2 L_-}{a_2^+ - a_2^-}, \quad \tilde{L}_3^2 p = M^2 L_1^2 p - L_2^2 p,$$

$$L_2 \tilde{L}_3 p = \frac{\tilde{L}_3 L_+ - \tilde{L}_3 L_-}{a_2^+ - a_2^-}, \quad L_1 \tilde{L}_3 p = \frac{\tilde{L}_3 L_+ - a_2^+ L_2 \tilde{L}_3 p}{a_1},$$

and the aggregates

$$\hat{T}^2 p,\ \hat{T}\hat{\xi}p,\ \hat{T}\hat{\eta}p,\ \hat{\xi}^2 p,\ \hat{\xi}\hat{\eta}p,\ \hat{\eta}^2 p$$

can be estimated with the help (6.21).

And, finally, we write down the obvious relations:

$$\hat{\tau}p - \hat{\xi}p + \omega\hat{\eta}p - Tp + \hat{\xi}p = 0,$$

$$\hat{\tau}(Tp) - \hat{\xi}(Tp) + \omega\hat{\eta}(\hat{T}p) - \hat{T}^2 p + \hat{T}\hat{\xi}p = 0,$$

$$\hat{\tau}(\hat{\xi}p) - \hat{\xi}(\hat{\xi}p) + \omega\hat{\eta}(\hat{\xi}p) - \hat{T}\hat{\xi}p + \hat{\xi}^2 p = 0,$$

$$\hat{\tau}(\hat{\eta}p) - \hat{\xi}(\hat{\eta}p) + \omega\hat{\eta}(\hat{\eta}p) - \hat{T}\hat{\eta}p + \hat{\xi}\hat{\eta}p = 0,$$

$$\hat{\tau}v - \hat{\xi}v + \omega\hat{\eta}v - 2\hat{T}v - \frac{1}{M^2}\hat{\eta}p = 0.$$

We introduce into consideration the aggregate

$$\hat{J}_k = \varepsilon_1 \{||\mathbf{Z}^k|| + ||\Omega^k||^2 + ||\hat{T}v^k||^2\} + ||\mathbf{X}^k||_D^2 +$$

$$+ ||p^k||^2 + ||\hat{T}p^k||^2 + ||\hat{\xi}p^k||^2 + ||\hat{\eta}p^k||^2 + ||v^k||^2,$$

where $\varepsilon_1 > 0$ is a constant. It follows from (6.19)–(6.22) that

$$\hat{J}_{k+1} - \hat{J}_k \le K_1 \Delta (J_k + J_{k+1}) \tag{6.23}$$

where

$$J_k = ||\mathbf{U}^k|| + ||\Omega^k||^2 + ||\hat{T}v^k||^2\} + ||\hat{T}p^k||^2 + ||\hat{\xi}p^k||^2 + ||\hat{\eta}p^k||^2 +$$

$$+||\hat{T}^2 p^k||^2 + ||\hat{T}\hat{\xi} p^k||^2 + ||\hat{T}\hat{\eta} p^k||^2 + ||\hat{\xi}^2 p^k||^2 + ||\hat{\xi}\hat{\eta} p^k||^2 + \hat{\eta}^2 p^k||^2,$$

$$||\mathbb{U}^k||^2 = h_y \sum_{j=-\infty}^{\infty} \int_{i=0}^{\infty} |\mathbb{U}_j^k(x)|^2 dx, \quad k_1 > 0$$

is a constant. According to the lemma from [13], the following estimation follows from (6.23)

$$J_k \leq \text{const} \cdot J_0, \quad k = 1, 2, \ldots, \tag{6.24}$$

and thus we establish the well-posedness of difference-differential **model I**. The inequality $\alpha > \delta$ is the sufficient condition for its stability.

Remark 6.5. The estimation analogous to (6.24) can be easily obtained for **model II**.

Remark 6.6. A more accurate estimation can be obtained (see [4])

$$G_k \leq \text{const} \cdot G_0, \quad k = 1, 2, \ldots, \tag{6.25}$$

where

$$G_k = ||\mathbb{U}^k|| + ||\hat{T}\mathbb{U}^k||^2 + ||\hat{\xi}\mathbb{U}^k||^2 + ||\hat{\eta}\mathbb{U}^k||^2 +$$

$$+||\hat{T}^2\mathbb{U}^k||^2 + ||\hat{T}\hat{\xi}\mathbb{U}^k||^2 + ||\hat{T}\hat{\eta}\mathbb{U}^k||^2 + ||\hat{\xi}^2\mathbb{U}^k||^2 + ||\hat{\xi}\hat{\eta}\mathbb{U}^k||^2 + ||\hat{\eta}^2\mathbb{U}^k||^2,$$

$$||\hat{T}\mathbb{U}^k||^2 = h_y \sum_{j=-\infty}^{\infty} \int_0^{\infty} |\hat{T}\mathbb{U}_j^k(x)|^2 dx$$

and so on.

7. Solvability of difference-differential models

Estimation (6.24) from section 6 (precisely, estimation (6.25)) can be used to rpove existence of sufficiently smooth solution to initial **mixed problem II** ([4,8]). However, it needs to evaluate the aggregate G_0 by means of initial data. Consequently, there appears a problem on solvability of the suggested difference-differential models. We exemplify this problem by **model II**. After simple operations over (6.4), it is transformed into

$$B\xi\mathbb{V}^k + C\eta\mathbb{V}^k + \frac{1}{\alpha\Delta} \cdot A\mathbb{V}^k = -\frac{1}{\alpha} \cdot \Omega^k, \quad (x, y) \in R_+^2, \tag{7.1}$$

where

$$\mathbb{V}^k = (\varphi - 1)\mathbb{U}^k, \quad \Omega^k = B \cdot \xi\mathbb{U}^k + C\eta\mathbb{U}^k.$$

Boundary conditions (6.5) transforms into

$$S_0\mathbb{V}^k(0, y) = \Delta \cdot \mathbb{A}^k(0, y), \quad y \in R^1. \tag{7.2}$$

Here

$$S_0 = \begin{bmatrix} d & 1 & 0 \\ -\lambda \cdot \alpha \cdot \Delta \cdot \eta & 0 & 1 + \omega \cdot \alpha \cdot \Delta \cdot \eta \end{bmatrix},$$

$$\Lambda^k(0,y) = \{\lambda \cdot \eta \cdot p^k(0,y) - \omega \cdot \eta v^k(0,y)\} \cdot \begin{bmatrix} 0 \\ 1 \end{bmatrix}.$$

Applying the Fourier transform to (7.1) and (7.2), we obtain as the result a system of ordinary differential equations

$$B\xi \hat{\mathbf{V}}^k = -\left\{ i\xi C + \frac{1}{\alpha\Delta} \cdot A \right\} \hat{\mathbf{V}}^k - \frac{1}{\alpha}\hat{\Omega}^k, \tag{7.1$'$}$$

where

$$\hat{\mathbf{V}}^k = \hat{\mathbf{V}}(k\cdot\Delta, x, \zeta) = \frac{1}{\sqrt{2\pi}} \cdot \int_{R^1} e^{-i\zeta y} \mathbf{V}(k\Delta, x, y) dy,$$

$$\hat{\Omega}^k = \hat{\Omega}(k\cdot\Delta, x, \zeta) = \frac{1}{\sqrt{2\pi}} \cdot \int_{R^1} e^{-i\zeta y} \Omega(k\Delta, x, y) dy,$$

are the Fourier transforms of $\mathbf{V}$, Ω. Moreover, boundary conditions have the form

$$\tilde{S}_0 \hat{\mathbf{V}}^k(0,\zeta) = \Delta \cdot \hat{\Lambda}^k(0,\zeta), \tag{7.2$'$}$$

Here

$$\tilde{S}_0 = \begin{bmatrix} d & 1 & 0 \\ -\lambda\cdot\tilde{\zeta}\cdot i & 0 & 1+\omega\cdot\tilde{\zeta}\cdot i \end{bmatrix},$$

$$\hat{\mathbf{V}}^k(0,\zeta) = \frac{1}{\sqrt{2\pi}} \cdot \int_{R^1} e^{-i\zeta y} \mathbf{V}(k\cdot\Delta, 0, y) dy,$$

$$\hat{\Lambda}(0,\zeta) = \frac{1}{\sqrt{2\pi}} \int_{R^1} e^{-i\zeta y} \Lambda(k\cdot\Delta, 0, y) dy, \tilde{\zeta} = \alpha\Delta\zeta.$$

Multiplying (7.1$'$) from the left by B^{-1}, we reduce it to the form

$$\xi \hat{\mathbf{V}}^k = \mathcal{M} \cdot \hat{\mathbf{V}}^k + \Psi, \tag{7.1$''$}$$

where

$$\mathcal{M} = \begin{bmatrix} \frac{M^2\tilde{\kappa}}{\beta^2} & -\frac{M^2\tilde{\kappa}}{\beta^2} & i\zeta\frac{M^2}{\beta^2} \\ -\frac{\tilde{\kappa}}{\beta^2} & \frac{M^2\tilde{\kappa}}{\beta^2} & -i\zeta\frac{1}{\beta^2} \\ -\frac{i\zeta}{M^2} & 0 & -\tilde{\kappa} \end{bmatrix},$$

$$B^{-1} = \begin{bmatrix} -\frac{M^2}{\beta^2} & \frac{1}{\beta^2} & 0 \\ \frac{1}{\beta^2} & -\frac{1}{\beta^2} & 0 \\ 0 & 0 & \tilde{\kappa}^2 \end{bmatrix},$$

$$\tilde{\kappa} = \frac{1}{\alpha\Delta} + i\zeta\omega, \quad \Psi = -\frac{1}{\alpha}\{\xi\hat{\mathbb{U}}^k + i\tilde{\zeta}B^{-1}C\hat{\mathbb{U}}^k\}.$$

We now determine eigen-values of $\mathcal{M}$:

$$\lambda_{1,2}(\mathcal{M}) = \frac{Ms \pm \sigma}{\alpha\Delta\beta^2}, \quad \lambda_3(\mathcal{M}) = -\tilde{\kappa},$$

where $s = M + i\zeta\omega M = M + i\chi$, $\sigma = \sqrt{s^2 + w^2}$, $w = \tilde{\zeta}\beta$, $\chi = \tilde{\zeta}\omega M$. We note that

$$\operatorname{Re}\sigma = \left(\frac{((M^2 - \chi^2 + w^2)^2 + 4M^2\chi^2)^{1/2} + M^2 - \chi^2 + w^2}{2}\right)^{1/2} > M > 0$$

for any χ, w. Hence

$$\operatorname{Re}\lambda_1(\mathcal{M}) > 0, \quad \operatorname{Re}\lambda_2(\mathcal{M}) < 0.$$

We present $\mathcal{M}$ in the form:

$$\mathcal{M} = A^{-1/2}F^{-1}DFA^{1/2}$$

where

$$A^{1/2} = \operatorname{diag}(1, M, M), \quad A^{-1/2} = \operatorname{diag}(1, \kappa, \kappa),$$
$$D = \operatorname{diag}(\lambda_1, \lambda_2, \lambda_3), \quad F_2 = \operatorname{diag}(s - M\sigma, s + M\sigma, 1)$$

are diagonal matrices;

$$F = \frac{1}{s^2 - M^2\tilde{\zeta}^2}F_2F_1$$

$$F_1 = \begin{bmatrix} \frac{1}{2\beta^2} & -\frac{s}{2\sigma\beta^2} & \frac{i\tilde{\zeta}M}{2\sigma\beta^2} \\ \frac{1}{2\beta^2} & \frac{s}{2\sigma\beta^2} & -\frac{i\tilde{\zeta}M}{2\sigma\beta^2} \\ i\tilde{\zeta} & -i\tilde{\zeta}M & s \end{bmatrix},$$

$$F^{-1} = \begin{bmatrix} s + M\sigma & s - M\sigma & 0 \\ -\sigma - Ms & \sigma - Ms & i\tilde{\zeta}M \\ -i\tilde{\zeta}\beta^2 & -i\tilde{\zeta}\beta^2 & s \end{bmatrix}.$$

We note that $s^2 - M^2\tilde{\zeta}^2 \neq 0$ for every $\zeta \in R_1$ if $\omega \neq 0$.

We continue the vector function Ψ by zero at $x < 0$. As far as for all $\zeta \in R^1$ $\operatorname{Re}\lambda_{1,2,3}(\mathcal{M}) \neq 0$ the unique bounded solution to (7.1″) at $x \in R^1$ can be constructed:

$$\mathbb{Y}(x,\zeta) = \int_{R^1} H(x - \rho, \zeta) \cdot \tilde{\Psi}(\rho,\zeta)d\rho.$$

Here $H(\rho,\zeta)$ is the Green matrix given by the formula

$$H(\rho,\zeta) = \begin{cases} A^{-1/2}F^{-1}D_-FA^{1/2} \text{ at } \rho > 0, \\ -A^{-1/2}F^{-1}D_+FA^{1/2} \text{ at } \rho < 0. \end{cases}$$

$$D_+ = \mathrm{diag}(\exp(\rho \cdot \lambda_1(\mathcal{M})), 0, 0),$$

$$D_- = \mathrm{diag}(0, \exp(\rho \cdot \lambda_2(\mathcal{M})), \exp(\rho \cdot \lambda_3(\mathcal{M}))),$$

are diagonal matrices;

$$\tilde{\mathbb{\Psi}}(\rho,\zeta) = \begin{cases} \mathbb{\Psi}(\rho,\zeta) & \text{at } \rho \geq 0, \\ 0 & \text{at } \rho < 0. \end{cases}$$

Using the vector function $\mathbb{Y}(x,\zeta)$, we seek the solution to (7.1″), (7.2′) in the form

$$\hat{\mathbb{V}}^k(x,\zeta) = \mathbb{q}(x,\zeta) + \mathbb{Y}(x,\zeta),$$

where $\mathbb{q}$ is the solution to the boundary problem

$$\xi\mathbb{q} = \mathcal{M} \cdot \mathbb{q}, \quad x > 0, \tag{7.3}$$

$$\tilde{S}_0\mathbb{q}(0,\zeta) = \Delta \cdot \hat{\mathbb{A}}^k(0,\zeta) - \tilde{S}_0\mathbb{Y}(0,\zeta), \quad x = 0. \tag{7.4}$$

It follows from (7.3) that

$$\mathbb{q} = \mathbb{q}(x,\zeta) = A^{-1/2}F^{-1}e^{xD}FA^{1/2} \cdot \mathbb{q}(0,\zeta) =$$

$$= A^{-1/2}F^{-1} \cdot \begin{bmatrix} \exp[\lambda_1(\mathcal{M}) \cdot x] \cdot C^1 \\ \exp[\lambda_2(\mathcal{M}) \cdot x] \cdot C^2 \\ \exp[\lambda_3(\mathcal{M}) \cdot x] \cdot C^3 \end{bmatrix},$$

where

$$\begin{pmatrix} C^1 \\ C^2 \\ C^3 \end{pmatrix} = FA^{1/2} \cdot \mathbb{q}(0,\zeta).$$

We note that the requirement on boundedness of $\mathbb{q}$ for all $x \geq 0$ is equivalent to $C^1 = 0$. So,

$$\mathbb{q} = A^{-1/2}F^{-1} \cdot \begin{pmatrix} 0 \\ \exp(\lambda_2(\mathcal{M}) \cdot x) \cdot C^2 \\ \exp(\lambda_3(\mathcal{M}) \cdot x) \cdot C^3 \end{pmatrix}.$$

Substituting this expression into (7.4), we derive that

$$\begin{pmatrix} C^2 \\ C^3 \end{pmatrix} = \hat{S}_0^{-1} \cdot \{\Delta \cdot \hat{\mathbb{A}}^k(0,\zeta) - \tilde{S}_0\mathbb{Y}(0,\zeta)\}$$

where

$$\hat{S}_0 = \begin{pmatrix} s(d-1) + \dfrac{\sigma(1-dM^2)}{M} & i\tilde{\zeta} \\ -i\tilde{\zeta}M(\lambda(\hat{\kappa}-\sigma) + \hat{\kappa}\beta^2\kappa^2 & \hat{\kappa}^2 \end{pmatrix}, \quad \hat{\kappa} = \alpha\Delta\tilde{\kappa}.$$

Thus, if the **Lopatinskii condition** holds for all $\zeta \in R^1$

$$L = \det \hat{S}_0 \neq 0, \tag{7.5}$$

then problem (7.3), (7.4) (together with problem (7.1″), (7.2)) is uniquely solvable.

We estimate now the obtained solution. Using the Minkovskii inequality, we easily come to

$$\left(\int_0^\infty |\hat{\mathbf{V}}^k|^2 dx\right)^{1/2} \leq \left(\int_0^\infty |\mathbf{q}|^2 dx\right)^{1/2} + \left(\int_0^\infty |\mathbf{Y}|^2 dx\right)^{1/2} \leq$$

$$\leq \left(\int_0^\infty |\mathbf{q}|^2 dx\right)^{1/2} + \left(\int_R |\mathbf{Y}|^2 dx\right)^{1/2}, \tag{7.6}$$

where $|\hat{\mathbf{V}}^k|^2 = (\hat{\mathbf{V}}^k, \overline{\hat{\mathbf{V}}}^k)$ and so on.

We estimate summonds from the right part of inequality (7.6). With the use of the known integral inequalities, we have

$$\left(\int_{R^1} |\mathbf{Y}|^2 dx\right)^{1/2} \leq \int_{R^1} ||H(x,\zeta)|| dx \cdot \left(\int_{R^1} |\tilde{\mathbf{\Psi}}|^2 dx\right)^{1/2} \leq$$

$$\leq \frac{\kappa}{\alpha} ||F^{-1}|| \cdot ||F|| \cdot ||B^{-1}|| \cdot \left\{ \frac{1}{\mathrm{Re}\lambda_1(\mathcal{M})} - \frac{2}{\mathrm{Re}[\lambda_2(\mathcal{M}) + \lambda_3(\mathcal{M})]} \right\} \times$$

$$\times \left(\int_0^\infty |\hat{\mathbf{\Omega}}|^2 dx\right)^{1/2}, \tag{7.7}$$

where $||F||$ is the operator norm of the matrix F and so on. Then, the following estimation are easy to check:

$$|\mathbf{Y}(0,\zeta)| = \left| \int_{R^1} H(-\rho,\zeta) \cdot \tilde{\mathbf{\Psi}}(\rho,\zeta) d\rho \right| \leq$$

$$\leq \kappa \frac{1}{\alpha\sqrt{2\mathrm{Re}\lambda_1(\mathcal{M})}} ||F^{-1}|| \cdot ||F|| \cdot ||B^{-1}|| \cdot \left(\int_0^\infty |\hat{\mathbf{\Omega}}|^2 dx\right)^{1/2},$$

$$|\mathbf{q}(x,\zeta)| \leq \kappa ||F^{-1}|| \cdot ||\hat{S}_0^{-1}|| \cdot \exp\left\{\frac{1}{2}\mathrm{Re}[\lambda_2(\mathcal{M}) + \lambda_3(\mathcal{M})]\right\} \times$$

$$\times \{\Delta \cdot |\hat{\mathbf{\Lambda}}^k(0,\zeta)| + ||\tilde{S}_0|| \cdot |\mathbf{Y}(0,\zeta)|\}.$$

From here

$$\left(\int_0^\infty |\mathbf{q}(x,\zeta)|^2 dx\right)^{1/2} \leq \frac{\kappa\|F^{-1}\|\cdot\|\hat{S}_0^{-1}\|}{\sqrt{|\mathrm{Re}(\lambda_2(\mathcal{M})+\lambda_3(\mathcal{M}))|}}\cdot\Bigg\{\Delta\cdot|\hat{\mathbb{A}}^k(0,\zeta)|+$$

$$+\|\tilde{S}_0\|\cdot\frac{\kappa}{\alpha\cdot\sqrt{2\mathrm{Re}\lambda_1(\mathcal{M})}}\|F^{-1}\|\cdot\|F\|\cdot\|B^{-1}\|\cdot\left(\int_0^\infty |\hat{\Omega}^k|^2 dx\right)^{1/2}\Bigg\}. \tag{7.8}$$

We estimate now the constants which enter (7.7), (7.8). Since $-M^2+\mathrm{Re}\sigma+\beta^2\neq 0$ for all $\zeta\in R^1$

$$\frac{1}{\mathrm{Re}\lambda_1(\mathcal{M})}-\frac{2}{\mathrm{Re}(\lambda_2(\mathcal{M})+\lambda_3(\mathcal{M}))}=$$

$$=\alpha\beta^2\Delta\frac{3\mathrm{Re}\sigma+1}{(M^2+\mathrm{Re}\sigma)(\beta^2+\mathrm{Re}\sigma-M^2)}\leq q_1\cdot\frac{\Delta}{A_1|\tilde{\zeta}|+B_1},$$

$$\frac{1}{2\mathrm{Re}\lambda_1(\mathcal{M})}=\frac{\alpha\beta^2\Delta}{2(M^2+\mathrm{Re}\sigma)}\leq q_2\cdot\Delta, \tag{7.9}$$

$$\|F^{-1}\|\cdot\|F\|\leq A_3|\tilde{\zeta}|+B_3,$$

$$\frac{1}{|\mathrm{Re}(\lambda_2(\mathcal{M})+\lambda_3(\mathcal{M}))|}=\frac{\alpha\beta^2\Delta}{(\beta^2+\mathrm{Re}\sigma-M^2)}\leq q_3\cdot\frac{\Delta}{A_2|\tilde{\zeta}|+B_2},$$

$$\|\hat{S}_0^{-1}\|\leq\frac{1}{A_4|\tilde{\zeta}|+B_4},\ \|\tilde{S}_0\|\leq A_5\cdot|\tilde{\zeta}|+B_5,$$

where $\tilde{\zeta}=\alpha\cdot\Delta\cdot\zeta$, $q_{1,2,3}, A_l, B_l>0, l=\overline{1,5}$ are constants independent of ζ. It follows from (3.6)–(3.9) that

$$\left(\int_0^\infty |\hat{\mathbb{V}}^k|^2 dx\right)^{1/2} \leq c_1\cdot\Delta^{3/2}\cdot|\hat{\mathbb{A}}^k(0,\zeta)|+$$

$$+\Delta\cdot(c_2+c_3\cdot\Delta\cdot|\zeta|)\cdot\left(\int_0^\infty |\hat{\Omega}^k(x,\zeta)|^2 dx\right)^{1/2},$$

where $c_{1,2,3}>0$ are constants. The Parceval equality applied to the squared inequality from above implies the desired estimation:

$$\|\mathbb{V}^k\|^2_{L_2(R^2_+)}=\int\int_{R^2_+}|\mathbb{V}^k(x,y)|^2 dxdy=\|\hat{\mathbb{V}}^k\|^2_{L_2(R^2_+)}=\int_{R^2_+}|\hat{\mathbb{V}}^k(x,\zeta)|^2 dxd\zeta\leq$$

$$\leq 3c_1^2\cdot\Delta^3\cdot\|\mathbb{A}^k(0)\|^2_{L_2(R)}+3c_2^2\cdot\Delta^2\cdot\|\Omega^k\|^2_{L_2(R^2_+)}+3c_3^2\cdot\Delta^4\cdot\|\eta\Omega^k\|^2_{L_2(R^2_+)}. \tag{7.10}$$

Here

$$\|\mathbb{A}^k(0)\|^2_{L_2(R)} = \int\limits_{R^1} |\mathbb{A}^k(0,y)|^2 dy = \int\limits_{R^1} |\hat{\mathbb{A}}^k(0,\zeta)|^2 d\zeta,$$

$$\|\Omega^k\|^2_{L_2(R^2_+)} = \int\limits_{R^2_+}\int |\Omega^k(x,y)|^2 dx dy = \int\limits_{R^2_+} |\hat{\Omega}^k(x,\zeta)|^2 dx d\zeta$$

and so on.

As far as

$$\|\mathbb{U}^{k+1}\|_{L_2(R^2_+)} \le \|\mathbb{V}^k\|_{L_2(R^2_+)} + \|\mathbb{U}^k\|_{L_2(R^2_+)},$$

$\|\mathbb{U}^1\|_{L_2(R^2_+)}$ can be estimated in norms of the intial data with the help of (7.10). Dividing both parts of (7.10) by Δ^2, we estimate $\|\hat{\tau}\mathbb{U}^0\|_{L_2(R^2_+)}$ through the initial data. Reasoning in a similar way, we estimate the norms which enter G_0.

Remark 7.1. An analogous result holds true for **model I**. Indead, we rewrite (6.1) as follows:

$$B\xi\mathbb{V}^k + C\hat{\eta}_0\mathbb{V}_k + \frac{1}{\alpha\Delta}A\mathbb{V}^k = -\frac{1}{\alpha}\Omega^k, \quad x > 0,\ |j| = 0, 1, \cdots, \tag{7.11}$$

where

$$\mathbb{V}^k = \mathbb{V}^k(x,y) = \mathbb{U}^{k+1}_j(x) - \mathbb{U}^k_j(x), \quad \mathbb{U}^k = \mathbb{U}^k(x,y) = \mathbb{U}^k_j(x),$$

$$\Omega^k = B\xi\mathbb{U}^k + C\hat{\eta}_0\mathbb{U}^k, \quad y = jh_y.$$

Boundary condition (6.2) transform into:

$$S_0\mathbb{V}^k(0,y) = \Delta\cdot\mathbb{A}^k(0,y), \quad |j| = 0, 1, \cdots. \tag{7.12}$$

Here

$$S_0 = \begin{bmatrix} d & 1 & 0 \\ -\lambda\cdot\alpha\cdot\Delta\cdot\hat{\eta}_0 & 0 & 1+\omega\cdot\alpha\cdot\Delta\cdot\hat{\eta}_0 \end{bmatrix},$$

$$\Lambda^k(0,y) = \{\lambda\cdot\hat{\eta}_0\cdot p^k(0,y) - \omega\cdot\hat{\eta}_0\mathbb{V}^k(0,y)\}\cdot\begin{bmatrix} 0 \\ 1 \end{bmatrix}.$$

Apllying the Fourier transform to (7.11), (7.12), we obtain the boundary-value problem for the ordinary differential equations system:

$$B\xi\hat{\mathbb{V}}^k = -\left\{i\hat{\xi}C + \frac{1}{\alpha\Delta}\cdot A\right\}\hat{\mathbb{V}}^k - \frac{1}{\alpha}\hat{\Omega}^k, \tag{7.11'}$$

$$\tilde{S}_0\hat{\mathbb{V}}^k(0,\zeta) = \Delta\cdot\hat{\Lambda}^k(0,\zeta). \tag{7.12'}$$

Here

$$\tilde{S}_0 = \begin{bmatrix} d & 1 & 0 \\ -\lambda\cdot\tilde{\zeta}\cdot i & 0 & 1+\omega\cdot\tilde{\zeta}\cdot i \end{bmatrix},$$

$$\hat{\mathbb{V}}^k(x,\zeta) = \frac{1}{\sqrt{2\pi}} \cdot \int\limits_{R^1} e^{-i\zeta y}\mathbb{V}^k(x,y)dy, \quad \text{and so on,}$$

$$\tilde{\zeta} = \alpha\Delta\hat{\zeta}, \quad \hat{\zeta} = \frac{\sin(h_y \cdot \zeta)}{h_y}.$$

Then, consider $\frac{h_y}{\Delta}$ constant,for problem (7.11′)–(7.12′) we obtain the estimation

$$\left(\int\limits_0^\infty |\hat{\mathbb{V}}^k(0,\zeta)|^2 dx\right)^{1/2} \le c_1 \cdot \Delta \cdot |\hat{\mathbb{A}}^k(0,\zeta)| + \Delta \cdot c_2 \cdot \left(\int\limits_0^\infty |\hat{\Omega}^k(x,\zeta)|^2 dx\right)^{1/2},$$

where $c_{1,2} > 0$ are constants independent of ζ. Finally we have:

$$||\mathbb{V}^k||^2 \le 2c_1^2\Delta^2 \cdot ||\mathbb{A}^k(0)||^2 + 2c_2^2\Delta^2 \cdot ||\Omega||^2.$$

Here

$$||\mathbb{V}^k||^2 = h_y \sum_{j=-\infty}^{\infty} \int\limits_0^\infty |\hat{\mathbb{V}}_j^k(x)|^2 dx$$

$$||\mathbb{A}(0)||^2 = h_y \sum_{j=-\infty}^{\infty} |\mathbb{A}_j^k(0)|^2$$

and so on.

Remark 7.2. In secitons 6, 7 the authors used results from papers [17,18].

8. Spline-schemes for problem I

We will assume that $0 < x < l < \infty$ in **problem 2I** and by this reason we seek the solution to **problem I** which satisfies boundary conditions (0.2) and, additionally, boundary conditions at $x = l$:

$$\mathbb{U}^{II} = R \cdot \mathbb{U}^I,$$

where R is a real constant matrix of dimension $N_1 \times N_0$. For **problem I** formulated in such way we consider the following **difference-differential model**:

$$A \cdot \hat{\mathbb{U}} = A \cdot L_y \mathbb{U} - \Delta \cdot B \cdot \hat{\xi}\mathbb{U} - \frac{1}{2} r_y \cdot C \cdot \eta_0 \mathbb{U}, \tag{8.1}$$

$$0 < x < l, \quad k, |j| = 0, 1, \dots,$$

$$\mathbb{U}^I = S \cdot \mathbb{U}^{II}, \quad x = 0, \ k, |j| = 0, 1, \dots, \tag{8.2}$$

$$\mathbb{U}^{II} = R \cdot \mathbb{U}^I, \quad x = l, \ k, |j| = 0, 1, \dots, \tag{8.3}$$

$$\mathbb{U}_j^0(x) = \mathbb{U}_0(x, j \cdot h_y), \quad 0 < x < l, \ |j| = 0, 1, \dots. \tag{8.4}$$

Here

$$\mathbf{U} = \mathbf{U}_j^k(x), \quad \hat{\mathbf{U}} = \mathbf{U}_j^{k+1}(x), \quad \hat{\xi}\mathbf{U} = \frac{\partial}{\partial x}\hat{\mathbf{U}},$$

$$L_y = \frac{\theta + \theta^{-1}}{2}, \quad \eta_0 = \frac{\theta - \theta^{-1}}{2}.$$

Then, the vector function $\mathbf{U}_j^k(x)$ at fixed k, j is sought in the form of **cubic interpolational spline of deficiency 2** (the so-called **Hermitian cubic spline** , see [19]):

$$\mathbf{S}_3(x) = \left(1 - \frac{x - x_i}{h_x}\right)^2 \cdot \left(1 + 2 \cdot \frac{x - x_i}{h_x}\right) \cdot \mathbf{U}_j^k(x_i) +$$

$$+ \left(\frac{x - x_i}{h_x}\right)^2 \cdot \left(3 - 2 \cdot \frac{x - x_i}{h_x}\right) \cdot \mathbf{U}_j^k(x_{i+1}) +$$

$$+(x - x_i) \cdot \left(1 - \frac{x - x_i}{h_x}\right) \cdot \left[\mathbf{m}_i \cdot \left(1 - \frac{x - x_i}{h_x}\right) - \mathbf{m}_{i+1} \cdot \frac{x - x_i}{h_x}\right],$$

$$x \in [x_i, x_{i+1}], \quad i = \overline{0, n-1}, \quad x_0 = 0, \quad x_n = l, \quad l = nh_x, \tag{8.5}$$

where

$$x_i = i \cdot h_x, \quad \mathbf{m}_i = \frac{\partial}{\partial x}\mathbf{U}_j^k(x_i).$$

According to [19], we require fulfilment of the relations:

$$\mathbf{S}_3''(x_i + 0) - \mathbf{S}_3''(x_i - 0) = \alpha_i(\cdot \mathbf{S}_3'''(x_i + 0) - \mathbf{S}_3'''(x_i - 0)),$$

$$i = \overline{1, n-1}, \tag{8.6}$$

Here α_i are given numbers.

Substituting (8.5) into (8.6), we arrive at

$$(1 - 3\bar{\alpha}_i)\mathbf{m}_{i-1} + 4\mathbf{m}_i + (1 + 3\bar{\alpha}_i)\mathbf{m}_{i+1} =$$

$$= 3(1 + 2\bar{\alpha}_i) \cdot \frac{\mathbf{U}_{i+1,j}^k - \mathbf{U}_{ij}^k}{h_x} + 3(1 - 2\bar{\alpha}_i) \cdot \frac{\mathbf{U}_{i,j}^k - \mathbf{U}_{i-1,j}^k}{h_x},$$

$$i = \overline{1, n-1}. \tag{8.7}$$

where $\bar{\alpha}_i = \dfrac{\alpha_i}{h_x}$. We consider relations (8.7) at the $k+1$-th layer taking into account equations (8.1) at $x = x_i$:

$$[(1 - 3\bar{\alpha}_i)A + 3(1 - 2\bar{\alpha}_i)r_x B]\hat{\mathbf{m}}_{i-1} + (4A - 6\bar{\alpha}_i r_x B)\hat{\mathbf{m}}_i +$$

$$+[(1 + 3\bar{\alpha}_i)A + 3(1 + 2\bar{\alpha}_i)r_x B]\hat{\mathbf{m}}_{i+1} =$$

$$= 3(1 + 2\bar{\alpha}_i) \cdot \frac{\mathbf{Z}_{i+1}^k - \mathbf{Z}_i^k}{h_x} + 3(1 - 2\bar{\alpha}_i) \cdot \frac{\mathbf{Z}_i^k - \mathbf{Z}_{i-1}^k}{h_x},$$

$$i = \overline{1, n-1}. \tag{8.8}$$

Here

$$r_x = \frac{\Delta}{h_x},$$

$$\mathbf{Z}_i^k = A \cdot L_y \mathrm{U}_j^k(x_i) - r_y C \cdot \eta_0 \mathrm{U}_j^k(x_i), \quad \hat{\mathrm{m}}_i = \frac{\partial}{\partial x} \mathrm{U}_j^{k+1}(x_i).$$

We consider system (8.8) as a system of algebraic equations with respect to unknowns $\hat{\mathrm{m}}_i$, $i = \overline{0, n}$. For the system to be uniquely solvable it is necessary to pose also boundary conditions at $i = 0$ and $i = n$. The following relations will serve as boundary conditions:

$$\mathrm{S}_3'''(x_1 + 0) = \mathrm{S}_3'''(x_1 - 0),$$

$$\mathrm{S}_3'''(x_n + 0) = \mathrm{S}_3'''(x_n - 0).$$

which with account of (8.5), (8.1) can be rewritten in the form:

$$(A - 2r_x B)\hat{\mathrm{m}}_0 + 4r_x B\hat{\mathrm{m}}_1 - (A + 2r_x B)\hat{\mathrm{m}}_2 = 2 \cdot \frac{\mathbf{Z}_1^k - \mathbf{Z}_0^k}{h_x} - 2 \cdot \frac{\mathbf{Z}_2^k - \mathbf{Z}_1^k}{h_x},$$

$$(A + 2r_x B)\hat{\mathrm{m}}_n - 4r_x B\hat{\mathrm{m}}_{n-1} - (A - 2r_x B)\hat{\mathrm{m}}_{n-2} =$$

$$= 2 \cdot \frac{\mathbf{Z}_n^k - \mathbf{Z}_{n-1}^k}{h_x} - 2 \cdot \frac{\mathbf{Z}_{n-1}^k - \mathbf{Z}_{n-2}^k}{h_x}. \tag{8.9}$$

Apllying the matrix sweep method to (8.8), (8.9), we determine $\hat{\mathrm{m}}_i$, $i = \overline{0, n}$. Values of solutions $\mathrm{U}_j^{k+1}(x_i)$ are found from (8.1) and boundary conditions (8.2), (8.3). Parameters α_i, $i = \overline{1, n-1}$, r_x are chosen in order to ensure stability of the matrix sweep method.

Remark 8.1. For the case of dissipative boundary conditions (8.2), (8.3) it is easy to stability of difference-differential model (8.1)–(8.4) is easy to establish under the requirement

$$r_y \nu_c \leq \nu_0,$$

where ν_c, ν_0 are constants from section 1.

Remark 8.2. From the authors' point of view, the described method to find approximate solutions to difference-differential model (8.1)–(8.4) with the help of spline-functions is of great practical use.

9. Method of lines for gas dynamics equations

In this section a difference-differential scheme to obtain approximate solutions to gas dynamics equations (0.20) is considered. Finally it reduces to boundary-value problem for a system of differential equations (in concrete cases it is solved by the orthogonal sweep method).

Theoretical justification of this calcualtion model on the example of **mixed problem II** is given in sections 6, 7. The case of two bboundaries is investigated in [20].

Mostly this section is devoted to description of the calculation algorythm. As the example we take the problem on supersonic flowing of perfect gas around the cone. This stationary solution of gas dynamics equations is found by the establishment method. An approximate solution is found with the help of the suggested calculation model. Results are compared with the results of some other authors.

Following [4] we formulate the mixed problem: in the domain

$$t > 0, \quad 0 < s < 1, \quad 0 < \varphi < \pi$$

the solution to system

$$\begin{gathered}\frac{d}{dt}u - v^2 - w^2 = 0, \quad \frac{d}{dt}\Psi = 0, \\ \frac{d}{dt}v + \frac{c^2}{\varepsilon}P_s + uv - w^2\mathrm{ctg}\Theta = 0 \\ \frac{d}{dt}w - \frac{c^2}{\varepsilon}a_2P_s + \frac{c^2}{\sin\Theta}P_\varphi + uw + uv\mathrm{ctg}\Theta = 0 \\ \frac{d}{dt}P + \frac{1}{\varepsilon}v_s - \frac{a^2}{\varepsilon}w_s + \frac{1}{\sin\Theta}w_\varphi + 2u + w\mathrm{ctg}\Theta = 0,\end{gathered} \tag{9.1}$$

is sought which satisfies boundary conditions at $s = 1$ **(on the head shock wave)**:

$$\begin{gathered}u = u_\infty, \\ v = v_\infty + \frac{G(p)}{(1 + a_2^2(1))^{1/2}}, \\ w = w_\infty - a_2(1)\frac{G(p)}{(1 + a_2^2(1))^{1/2}}, \\ V = \frac{V_\infty(hp_\infty + p)}{(hp + p_\infty)}, \\ \frac{\partial}{\partial t}\Theta_s + w_\infty a_2(1) = v_\infty + \frac{G_1(p)}{(1 + a_2^2(1))^{1/2}},\end{gathered} \tag{9.2}$$

at $s = 0$ **(condition that no flows pass the boundary)**:

$$v = 0, \tag{9.3}$$

and the **symmetry conditions** at $\varphi = 0, \pi$:

$$w = P_\varphi = u_\varphi = v_\varphi = \Psi_\varphi = 0, \quad \frac{\partial}{\partial\varphi}\Theta_s = 0 \tag{9.4}$$

and **initial data** at $t = 0$:

$$u = u_0(s, \varphi), \quad v = v_0(s, \varphi), \quad w = w_0(s, \varphi),$$

$$P = P_0(s, \varphi), \quad \Psi = \Psi_0(s, \varphi), \quad \Theta_s = \Theta_{s0}(s, \varphi). \tag{9.5}$$

Here

$$\frac{d}{dt} = \frac{\partial}{\partial t} + \tilde{\Delta}\frac{\partial}{\partial s} + \frac{w}{\sin\Theta}\frac{\partial}{\partial \varphi}, \quad \tilde{\Delta} = \frac{a_1}{\varepsilon},$$

$$a_1 = v - s\frac{\partial}{\partial t}\Theta_s - wa_2, \quad a_2 = \frac{s}{\sin\Theta}\frac{\partial}{\partial\varphi}\Theta_s,$$

(R, Θ, φ) is the spherical coordinate system (in (9.1) $R = 1$); u, v, w are components of the velocity scaled by division by the maximal velocity $W_{\max}$;

$$P = \frac{1}{\gamma}\ln p, \quad \Psi = P + \ln V, \quad \gamma = 1.4,$$

p, ρ are the pressure and density of scaled by division by the bracing parameter p_T, ρ_T, $s = \dfrac{\Theta - \Theta_b}{\varepsilon}$, $\varepsilon = \Theta_s - \Theta_b$, Θ_b is the half-angle of the cone, $\Theta = \Theta_s(t, \varphi)$ is the equation of the shock wave on the sphere with radius $R = 1$, $c^2 = \dfrac{(\gamma - 1)\exp((\gamma - 1)P + \Psi)}{2}$ is the squared sound velocity,

$$u_\infty = U_\infty(\cos\alpha\cos\Theta_s - \sin\alpha\sin\Theta_s\cos\varphi),$$

$$v_\infty = -U_\infty(\sin\alpha\cos\Theta_s\cos\varphi + \cos\alpha\sin\Theta_s),$$

$$w_\infty = U_\infty\sin\alpha\sin\varphi,$$

α is the angle of attack (the vector of velocity filling flow lies in symmetry plane),

$$U_\infty = \left(\frac{(\gamma - 1)M_\infty^2/2}{1 + (\gamma - 1)M_\infty^2/2}\right)^{1/2},$$

$M_\infty > 1$ is the Mach number of the filling flow (see fig.1),

$$G(p) = (\hbar(p - p_\infty)(V_\infty - V))^{1/2},$$

$$G_1(p) = \frac{V_\infty G(p)}{V_\infty - V},$$

$$V = \frac{1}{\rho}, \quad V_\infty = \frac{1}{\rho_\infty}, \quad p_\infty = \rho_\infty^\gamma,$$

$$\rho_\infty = \left(\frac{U_\infty^2}{(\gamma - 1)M_\infty^2/2}\right)^{1/(\gamma - 1)}$$

$$a_2(1) = \frac{1}{\sin\Theta_s}\frac{\partial}{\partial\varphi}\Theta_s, \quad h = \frac{\gamma + 1}{\gamma - 1}, \quad \hbar = \frac{\gamma - 1}{2\gamma}.$$

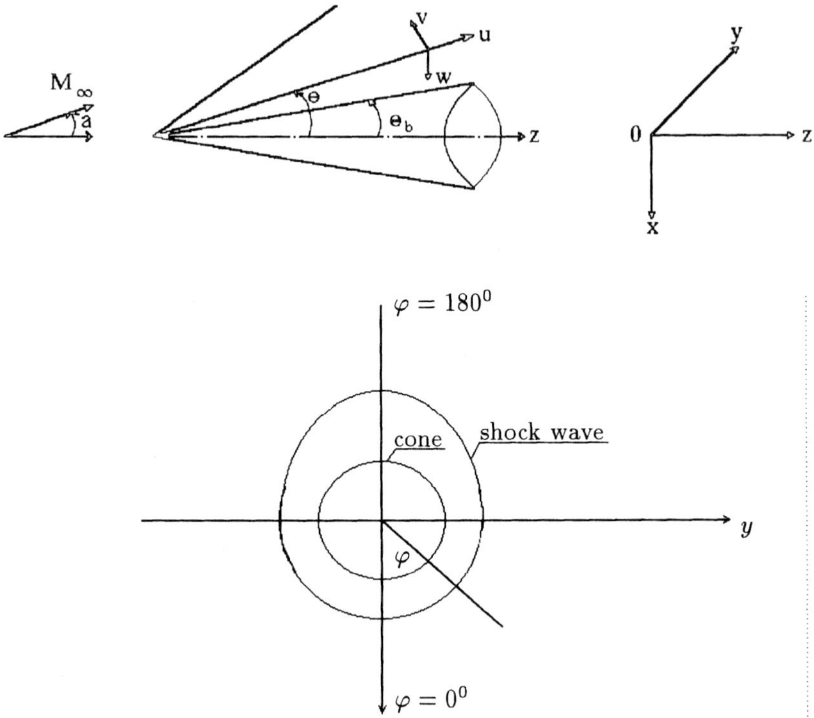

Fig 1.

We note also that it follows from the second and third relations from (9.2) that ($s = 1$):

$$a_2(1) = \frac{w_\infty - w}{v - v_\infty} \tag{9.6}$$

If $u, v, w, P, \Psi, \Theta_s$ do not depend on t, then problem (9.1)–(9.4) describes stationary supersonic flowing of the circular cone. In accordance with the establishment method this solution is sought y solving numerically (9.1)–(9.5).

The second, third and fourth equations of (9.1) are transfored into the system:

$$\mathbb{U}_s = M_1 \mathbb{U}_t + M_2 \mathbb{U}_\varphi + N, \tag{9.7}$$

$$\mathbb{U} = \begin{bmatrix} P \\ v \\ w \end{bmatrix}, \quad M_1 = -\frac{\varepsilon}{d} \begin{bmatrix} a_1 & -1 & a_2 \\ -c^2 & a_1 & 0 \\ c^2 a_2 & 0 & a_1 \end{bmatrix},$$

$$M_2 = -\frac{\varepsilon}{d\sin\Theta}\begin{bmatrix} b_1 & -w & b_2 \\ -c^2 w & b_1 & -c^2 \\ c^2 b_2 & -c^2 & b_1 \end{bmatrix},$$

$$N = -\frac{\varepsilon}{d}\begin{bmatrix} a_1F_1 - F_2 + F_3a_2 \\ -c^2F_1 + a_1F_2 + c^2a_2\Omega/\varepsilon \\ c^2a_2F_1 + a_1F_3 + c^2\Omega/\varepsilon \end{bmatrix},$$

$$d = a_1^2 - c^2a_2^2 - c^2, \quad b_1 = a_1w + c^2a_2, \quad b_2 = a_1 + a_2w,$$

$$F_1 = 2u + v\mathrm{ctg}\Theta, \quad F_2 = uv - w^2\mathrm{ctg}\Theta, \quad F_3 = uw + vw\mathrm{ctg}\Theta,$$

$$\Omega = a_2v_s + w_s - \frac{\varepsilon}{\sin\Theta}v_\varphi.$$

Following section 6, we carry out the discretization over t, φ:

$$\xi\mathbb{U}_j^{k+1}(s) = \frac{1}{\Delta}(M_1)_j^k(\mathbb{U}_j^{k+1}(s) - \mathbb{U}_j^k(s)) +$$

$$+\frac{1}{2h_\varphi}(M_2)_j^k(\mathbb{U}_{j+1}^{k+1}(s) - \mathbb{U}_{j-1}^{k+1}(s)) + (N)_j^k,$$

$$j = 1, 2, ..., J-1, k = 0, 1, \ldots. \tag{9.8}$$

Here $\xi = \frac{\partial}{\partial s}$, $\mathbb{U}_j^k(s) = \mathbb{U}(k\Delta, s, (j-1)h_\varphi)$, Δ, h_φ are disccretization steps; $\pi = (J-1)h_\varphi$. We note also that the index k in the parentheses means averaging of the sought functions which enter the entries of the matrices $M_{1,2}$, the vector N:

$$(\circ)^{(k)} = \frac{(\circ)^{k+1} + (\circ)^k}{2}$$

By nonlinearity of the initial problem, calculations at $k+1$-th layer actually must be carried out by iterations. At $\varphi = 0, \pi(j = 1, J) : w_j^k = 0$. We present the second and fourth equations from (9.1) at $\varphi = 0, \pi$ as a system :

$$(\mathbb{W}_j)_s = (M_0)_j(\mathbb{W}_j)_t + (N_0)_j,$$

and after discretization over t:

$$\xi\mathbb{W}_j^{k+1}(s) = \frac{1}{\Delta}(M_0)_j^k(\mathbb{W}_j^{k+1}(s) - \mathbb{W}_j^k(s)) + (N_0)_j^k,$$

$$j = 1, J, k = 0, 1, \ldots, \tag{9.9}$$

where

$$\mathbb{W} = \begin{bmatrix} P \\ v \end{bmatrix}, \quad \mathbb{W}_1^k(s) = \mathbb{W}(k\Delta, s, 0), \quad \mathbb{W}_J^k(s) = \mathbb{W}(k\Delta, s, \pi),$$

$$M_0 = \frac{\varepsilon}{c^2 - b^2}\begin{bmatrix} b & -1 \\ -c^2 & b \end{bmatrix}, \quad b = v - s\frac{\partial\Theta_s}{\partial t},$$

$$N_0 = \frac{\varepsilon}{c^2 - b^2} \begin{bmatrix} b(2u + v\mathrm{ctg}\Theta + \frac{1}{\sin\Theta} w_\varphi) - uv \\ \\ buv - c^2(2u + vctg\Theta + \frac{1}{\sin\Theta} w_\varphi) \end{bmatrix}.$$

Consequently, for the vector $\mathbb{Y}$:

$$\mathbb{Y} = \begin{bmatrix} \mathbb{W}_1^{k+1} \\ \mathbb{U}_2 k + 1 \\ \cdots \\ \mathbb{U}_{J-1}^{k+1} \\ \mathbb{W}_J^{k+1} \end{bmatrix}$$

we obtain a system of ordinary differential equations

$$\xi \mathbb{Y} = \tilde{A}\mathbb{Y} + \tilde{F}. \tag{9.10}$$

Here $\tilde{A}$ is a squre matrix of order $L = 3J-2$, $\tilde{F}$ is a vector of the same dimension. Elements of the matrix $\tilde{A}$ and the vector $\tilde{F}$ can be easily found via (9.8), (9.9).

We write down boundary conditions at $s = 0, 1$ for system (9.10). Thus, by virtue of (9.3), we have:

$$v_j^k(0) = 0, \quad j = 1, ..., J, \quad k = 0, 1, ...,$$

that is

$$\tilde{C}\mathbb{Y}(0) = 0, \tag{9.11}$$

where $\tilde{C}$ is a matrix of order $J \times L$,

$$\tilde{C} = \begin{bmatrix} 0 & 1 & 0 & 0 & \cdots & 0 & 0 & 0 & 0 & 0 \\ 0 & 0 & 0 & 1 & \cdots & 0 & 0 & 0 & 0 & 0 \\ \cdots & \cdots & \cdots & \cdots & \cdots & \cdots & \cdots & \cdots & \cdots & \cdots \\ 0 & 0 & 0 & 0 & \cdots & 0 & 1 & 0 & 0 & 0 \\ 0 & 0 & 0 & 0 & \cdots & 0 & 0 & 0 & 0 & 1 \end{bmatrix}$$

From (9.2) we arrive at

$$(\lambda_1)_j^k P_j^{k+1}(1) - v_j^{k+1}(1) = (\lambda_1)_j^k P_j^k(1) - v_j^k(1),$$

$$(\lambda_2)_j^k P_j^{k+1}(1) + w_j^{k+1}(1) = (\lambda_2)_j^k P_j^k(1) + w_j^k(1), \tag{9.12}$$

Here $j = 2, \cdots, J-1$,

$$\lambda_1 = \frac{\gamma P G'(p)}{\sqrt{1 + a_2^2(1)}}, \quad \lambda_2 = a_2(1)\lambda_1,$$

$$G'(p) = \hbar \frac{(3-\gamma)V_\infty + (\gamma+1)V}{4G_1(p)}$$

At $j = 1, J$ we have the following relation instead of (9.12):

$$(\lambda_0)_j^k P_j^{k+1}(1) - v_j^{k+1}(1) = (\lambda_0)_j^k P_j^k(1) - v_j^k(1), \qquad 9.13$$

where $\lambda_0 = \gamma p G'(p)$. In a view of 9.12), (9.13) we obtain

$$\tilde{B}\mathbb{Y}(1) = \tilde{\varphi}, \qquad (9.14)$$

where $\tilde{B}$ is a matrix of order $(2J-2) \times L$, $\tilde{\varphi}$ is of order $2J-2$. Elements of the matrix $\tilde{B}$ and the vector $\tilde{\varphi}$ can be found with the help of relations (9.12), (9.13). We note also that another ways to obtain boundary conditions at $s = 1$ for system (9.10) are also possible.

Formulated problem (9.10), (9.11), (9.14) for a system of ordinary differntial equations was solved then by the orthogonal sweep method [21].

The functions u and Ψ are found by the running count scheme.

$$\Psi_{ij}^{k+1} = \frac{\Psi_{ij}^k - d_1\Psi_{i+1,j}^{k+1} + d_2\Psi_{i,j-1}^{k+1}}{\tilde{d}}, \qquad i = 2, ..., I-1,\ j = 2, ..., J-1;$$

$$\Psi_{i,1}^{k+1} = \Psi_{I,1}^{k+1}, \qquad i = 1, ..., I-1;$$

$$\Psi_{1,j}^{k+1} = \Psi_{I,1}^{k+1}, \qquad j = 2, ..., J-1;$$

$$\Psi_{i,J}^{k+1} = \Psi_{I,J}^{k+1}, \qquad i = 1, ..., I-1;$$

$$u_{ij}^{k+1} = \frac{u_{ij}^k - d_1 u_{i+1,j}^{k+1} + d_2 u_{i,j-1}^{k+1} + \Delta[(v^2)_{(i),(j)}^{(k)} + (w^2)_{(i),(j)}^{(k)}]}{\tilde{d}},$$

$$i = 2, \ldots, I-1, \quad j = 2, ..., J-1;$$

$$u_{ij}^{k+1} = \frac{u_{ij}^k - \tilde{d}_1 u_{i+1,j}^{k+1} + \Delta(v^2)_{(i),(j)}^{(k)}}{1 - \tilde{d}_1},$$

$$i = 1, \ldots, I-1, \quad j = 1, J$$

Here

$$\Psi_{ij}^k = \Psi(k\Delta, (i-1)h_s, (j-1)h_\varphi), \quad i = \overline{1, I},$$

$$(I-1)h_s = 1, \quad d_1 = \frac{\Delta(\tilde{\Delta})_{(i)(j)}^{(k)}}{h_s}, \quad d_2 = \frac{\Delta(w/\sin\Theta)_{(i)(j)}^{(k)}}{h_\varphi},$$

$$\tilde{d} = 1 - d_1 + d_2; \quad \tilde{d}_1 = \Delta(b/\varepsilon)_{(i),j}, \quad j = 1, J$$

$$\Psi_{Ij}^k = P_{Ij}^k + \ln\frac{V_\infty(hp_\infty + p_{Ij}^k)}{(hp_{Ij}^k + p_\infty)}, \quad j = 1, \ldots, J;$$

$$u_{Ij}^{k+1} = (u_\infty)_{(j)}^{(k)}, \quad j = 1, \ldots, J;$$

$$(\circ)_{(i)} = \frac{(\circ)_{i+1} + (\circ)_i}{2}, \quad (\circ)_{(j)} = \frac{(\circ)_j + (\circ)_{j-1}}{2}.$$

Then, using the last relation from (9.2), we obtain $\Theta_s(t, \varphi)$:

$$(\Theta_s)_j^{k+1} = (\Theta_s)_j^k + \Delta\{v_\infty - w_\infty a_2(1) + G_1(p)\sqrt{1 + a_2^2(1)}\}_{(j)}^{(k)}.$$

Remark 9.1. Initial data at $\alpha = 0$ were given by simple analytic formulae. At $\alpha \neq 0$ the known functions for less values of α serve as initial data.

Numerical calculations were carried out by the algorythm from above. Since initial data were assigned roughly on the first steps we used iterations, then we calculated without iterations up to establishment. Establishment was considered to be reached if

$$\sum_{j=1}^{J}\left\{|(\Theta_s)_j^{k+1} - (\Theta_s)_j^k| + \left|\left(\frac{\partial}{\partial t}\Theta_s\right)_j^{k+1} - \left(\frac{\partial}{\partial t}\Theta_s\right)_j^k\right|\right\} < 10^{-6}$$

In calculations

$$M_\infty = 2,7; \quad \Theta_b = 10^\circ, 30^\circ.$$

At $\alpha = 0$, $M_\infty = 7$, $\Theta_b = 10^\circ$, 30° we reach establishment for 15 time steps, at $\alpha = 2$, $M_\infty = 2$, $\Theta_b = 10^\circ$, 30° for 30 time steps. The calculated values of parameters completely coincided with values from the table from [22].

With increasing of α a number of time steps also increased. Calculated parameters coincided with values from the table from [22] with high accuracy.

Remark 9.2. Results from [23] were used in this section [23].

References

1. Courant R., Friedrichs K., Levy H. On difference equations of mathematical physicsи// Advanteges in Mathematics (Uspehi mat.nauk) 1940. iss. 8. pp.125–160.

2. Godunov S.K. Equations of mathematical physics. Moscow, Nauka, 1979, 392 p.

3. Ladyzhenskaja O.A. Boundary problems of mathematical physics. Moscow, Nauka, 1973, 408 p.

4. Blokhin A.M. Energy integrals and their applications to problems of gas dynamics. Novosibirsk, Nauka, 1986, 240 p.

5. Volpert A.I., Hudyaev S.I. On Cauchy problem for composite systems of nonlinear differential equations. Mathematical collection (Matematicheskij sbornik), v. 87 (129), N. 4, 1972, pp. 504–528.

6. Blokhin A.M., Alajev R.D. On stability of modified difference scheme of McCormack for symmetric t-hyperbolic system. Nonclassical equations of mathematical physics (Neklassicheskije uravnenija matematicheskoj fiziki) Novosibirsk, Institute of mathematics of Russian Academy of Sciences, Siberian Branch, 1984, pp. 24–42.

7. Blokhin A.M., Alajev R.D., Druzhinin I.Yu. Stability of explicit differnce schemes for symmetric t-hyperbolic systems. Proceedings of Sobolev's seminar, Novosibirsk, Institute of mathematics of Russian Academy of Sciences, Siberian Branch, N. 1, 1986, pp. 26–39.

8. Blokhin A.M. Application of difference analogs of dissipative energy integrals to investigations of stability of difference schemes. Calculation problems in mathematical physics (Vychislitel'nyje problemy v zadachah mat. fiziki), Proceedings of Institute of mathematics of Russian Academy of Sciences, Siberian Branch, Novosibirsk, N. 11, 1988, pp. 67–93.

9. Godunov S.K., Ryaben'kij V.S. Introduction into theory of difference schemes. Fizmatgiz, 1962.

10. Blokhin A.M., Druzhinin I.Yu. Construction of difference examples of instability in problem on jump fitting. Simulation in mechanics, V. 2 (19), N. 3, 1988, pp. 22–31.

11. Blokhin A.M. Some topics from mathematical simulation of gas dynamics problems. Numerical analysis and Mathematical modeling, Banach Center Publications, V. 24, 1990, pp. 455–463.

12. Yanenko N.N. Method of fractional steps for solution of multidimensional problems of mathematical physics. Novosibirsk, Nauka, 1967, 196 p.

13. Rihtmajer R., Morton K. Difference methods for solution of boundary-value problems. Moscow, Mir, 1972, 420 p.

14. Blokhin A.M., Druzhinin I.Yu. Influence of boundary conditions on stability of certain difference splitting scheme. Embedding theorems and their applications to problems of mathematical physics, Novosibirsk, Institute of mathematics of USSR Academy of Sciences, Siberian Branch, 1989, pp. 38–47.

15. Bellman R. Introduction into matrix theory. Moscow, Nauka, 1976, 352 p.

16. Alajev R.D., Blokhin A.M. Theoretical and calculation investigation of difference model of mixed problem for gas dynamics equation with boundary conditions on shock wave. Nonclassical equations and equations of mixed type (Neklassicheskije uravnenija i uravnenija smeshannogo tipa) Novosibirsk, Institute of mathematics of USSR Academy of Sciences, Siberian Branch, 1983, pp. 66–83.

17. Blokhin A.M., Zimmerman V.R. Investigation of some difference-differential models for acoustics equations. Certain applications of functional analysis to problems of mathematical physics (Nekotoryje prilozhenija funkcional'-nogo analiza k problemam matematicheskoj fiziki), Novosibirsk, Institute of mathematics of USSR Academy of Sciences, Siberian Branch, 1988, pp. 21–32.

18. Blokhin A.M., Zimmerman V.R. Solvability of certain difference-differential models for acoustics equations. Embedding theorems and their applications to problems of mathematical physics (Teoremy vlozhenija i ih

prilozhenija k zadacham matematicheskoj fiziki), Novosibirsk, Institute of mathematics of USSR Academy of Sciences, Siberian Branch, 1989, pp. 48–59.

19. Zavjalov Yu.S., Kvasov B.I., Miroshnochenko V.L. Methods of spline-functions . Moscow, Nauka, 1980, 352 p.

20. Blokhin A.M., Zimmerman V.R. Investigation of difference-differntial model of linear mixed problem on supersonic flowing aroud wedge. Calculation problems in mathematical physics (Vychislitel'nyje problemy v matematicheskoj fizike), Proceedings of Institute of Mathematics of USSR Academy of Sciences, Siberian Branch. Novosibirsk, 1992.

21. Kuznetsov S.V. Solution of boundary problems for ordinary differntial equations. Calculations methods of linear algebra, Proceedings of Institute of mathematics of USSR Academy of Sciences, Siberian Branch, V. 6, 1985, pp. 85–110.

22. Babenko K.I, Voskresnskij G.P., Lubimov A.I., Rusanov V.V.Spatial flowing around smooth solids by perfect gasм. Moscow, Nauka, 1964, 505 p.

23. Blokhin A.M., Pozdeev A.A., Zimmerman V.R. Methods of lines for equations of gas dynamics: theoretical ground and calculation experiments. Calculation problems in mathematical physics (Vychislitel'nyje problemy v zadachah matematicheskoj fiziki), Proceedings of Institute of mathematics of USSR Academy of Sciences, Siberian Branch, V. 22, Novosibirsk, 1992.

24. Belyh V.N. Calculation algorythms without saturation in nonstationary problems of hydrodynamics in perfect liquid with free boundaries. Calculation problems in mathematical physics (Vychislitel'nyje problemy v zadachah matematicheskoj fiziki) Prtoceedings of Institute of mathematics of USSR Academy of Sciences, Siberian Branch, Novosibirsk, V. 11, 1988, pp. 3–67.

25. Vaskevich V.L. On approoximate solution of Dirichle problem in composite spatial domains. Numerical analysis (Chislennyj analiz), Proceedings of Institute of mathematics of USSR Academy of Sciences, Siberian Branch, Novosibirsk, V. 15, 1989, pp. 93–126.

Translated by E. V. Mishchenko

Chapter 6

Checking of a well-conditioning of an interval matrix

Ayche BULGAK[1]

Abstract. In this paper we investigate an algorithm for checking a well-conditioning of an interval matrix.

In this paper we investigate an algorithm for checking a well-conditioning of an interval matrix. Now we recall some definitions from [1], [2].

Definition 1. $\mathbf{A} = [\mathbf{A}_{ij}]$ is so-called interval matrix if it is $N \times N$ matrix and its entries are the real number intervals $\mathbf{A}_{ij} = [a_{ij}, b_{ij}]$.

Definition 2. An interval matrix $\mathbf{A}$ is said to be regular if every $A \in \mathbf{A}$ is a nonsingular matrix.

In [3] were given sufficient conditions for regularity of a matrix . But it is clear that in general case we cannot check directly the regularity of all matrices from interval matrix $\mathbf{A}$, since $\mathbf{A}$ is an uncountable set.

Definition 3. An interval matrix $\mathbf{A}^{-1} = [A^{-1}, A \in \mathbf{A}]$ is said to be an inverse interval matrix of $\mathbf{A}$.

Our aim is to suggest an algorithm for recognising an invertibility of a given $N \times N$ interval matrix $\mathbf{A}$.

It is well known, A is nonsingular matrix if and only if $\sigma_1(A)$, the minimal singular value of A, is greater then 0. There known good methods for computation the singular values of the matrices (see, for example [3]).

We cannot use this rule directly since in this case we will be work with uncountable set of $\mathbf{A}$.

Recall, the given problem is so-called well-conditioning if the small perturbations of elements lead to similar small change of solution.

In some cases very small perturbations lead to lose invertibility of given matrix.

This circumstance has led to new solvability criteria for equation $Ax = f$ (see, e.g., [3]). The most successful among them happened to be the test that uses the value $\mu(A) = ||A||||A^{-1}||$, here $||\cdot||$ is a spectral norm of a matrix. This value is so-called a condition number of matrix A.

[1]Research Center of Applied Mathematics, Selcuk University, Konya, Turkey

It is easily verified that if $\sigma_1(A) \neq 0$ then we have an equality

$$\mu(A) = ||A|| ||A^{-1}|| = \frac{\sigma_N(A)}{\sigma_1(A)}$$

where $\sigma_1(A)$, $\sigma_N(A)$ are respectively the minimal and maximal singular values of A.

The sensitivity of this value with respect to small perturbations is characterised by the

Theorem 1. ([3]) *If A is nonsingular matrix ($\mu(A) < \infty$) and*

$$2\mu(A)\frac{||B||}{||A||} < 1$$

then $A + B$ ($\mu(A + B) < \infty$) is also nonsingular and the following estimation

$$|\mu(A) - \mu(A + B)| \leq 3\mu^2(A)\frac{||B||}{||A||}$$

is true.

We can use the condition number $\mu(A) = ||A|| ||A^{-1}||$ for solving our problem.

Definition 4. A number $\mu(\mathbb{A}) = [\mu(A) = ||A|| ||A^{-1}||, A \in \mathbf{A}]$ is so-called a condition number of a given interval $N \times N$ matrix A.

The theorem 1 allows us to check an invertibility of the interval matrix $\mathbf{A}$ without computation $\mu(A)$ for all elements of $\mathbf{A}$. In this case we have a countable procedure, but unfortunately this procedure may be an infinity.

Example 1. Let

$$\mathbf{A} = \begin{bmatrix} [1,2] & 3 \\ 2 & 4 \end{bmatrix}$$

be 2×2 interval matrix. It is clear that this matrix is singular since

$$A = \begin{bmatrix} 1.5 & 3 \\ 2 & 4 \end{bmatrix}$$

is an element of the given set. For beginning we can compute a condition number of a matrix

$$A_0 = \begin{bmatrix} 1 & 3 \\ 2 & 4 \end{bmatrix}$$

from a given set. This allows us to write that

$$||A_0|| = 5.47, \quad \mu(A_0) = 14.945.$$

If we choose a value $\beta_1 = ||A_0||/(4\mu(A_0)) \approx 0.09$ then based on the theorem 1 we can conclude that the interval matrix ($0 < \beta < \beta_1$)

$$\mathbf{A}_1 = \begin{bmatrix} [1, 1+\beta] & 3 \\ 2 & 4 \end{bmatrix}$$

is nonsingular and $\mu(A_1) < 2\mu(A_0) < 30$. We can repeat our procedure for matrix

$$A_1 = \begin{bmatrix} [1, 1+\beta_1] & 3 \\ 2 & 4 \end{bmatrix}$$

and compute a value β_2 such that that the interval matrix $(0 < \beta < \beta_2)$

$$\mathbb{A}_2 = \begin{bmatrix} [1, 1+\beta_1+\beta] & 3 \\ 2 & 4 \end{bmatrix}$$

is nonsingular.

As a result we can compute a positive number sequence $\beta_1, \beta_2, \ldots, \beta_n, \ldots$,

$$\sum_{m=1}^{\infty} \beta_m < 0.5, \quad \beta_n \to 0, \quad n \to \infty$$

and a matrix sequence

$$A_n = A_{n-1} + \begin{bmatrix} \beta_n & 0 \\ 0 & 0 \end{bmatrix}, \quad \mu(A_n) \to \infty, \quad n \to \infty.$$

This procedure can't be interpretate as an algorithm, but it may be leads us to an algorithm. For this we must define a boundary between well-condition and ill-condition matrices.

There are many different way to express the degree of ill-conditioning the invertability of a matrix A. One of them is given by so called the parameter of practical invertability (we note μ^*).

We call the matrix A practical invertible (μ^*-invertible) if $A + B$ is also invertible for any matrix B such that $||B|| < ||A||/\mu^*$.

It is clear that an inequality $\mu(A) < \mu^*$ quarantes that matrix A is practically invertible, where $\mu(A)$ is the condition number.

It is naturally, the users choose the value μ^* in view of their concrete physical problem. Since , as a rule, we don't know exact the matrix A elements, we know them with some indefiniteness, which is measured by perturb matrix δA, $||\delta A||/||A|| < \varepsilon$, ε is given small positive number. Hence we can choose μ^* by notation $\mu^* = 2/\varepsilon$.

In this case, the theorem 1 allows us to conclude that for μ^*-invertible matrix A all close to it matrices $A + \delta(A)(||\delta A||/||A|| < 1/(2\mu^*)$ are $1.1\mu^*$-invertible.

Another way for choosing μ^* is connected with the algorithm which used for computation $Ax = f$ and depends on the particular computer involved.

Introduction of the value μ^* allows to organise the process of computation $Ax = f$ such that either compute it with all true digits (without last digit of its computer representation) or detect the ill-conditions of the problem, which is given by inequality $\mu(A) > \mu^*$.

From theorem 1 we have

Theorem 2. *If A is practically invertable ($\mu(A) < \mu^*$) and $\dfrac{||B||}{||A||} < \dfrac{1}{4\mu^*}$ then an estimation $\mu(A+B) < 1.5\mu^*$ is true.*

Now we present an algorithm for checking practically invertibility of a given interval matrix. We present it for special interval matrix

$$\mathbb{A} = \begin{bmatrix} [a_{11}, b_{11}] & a_{12} & a_{13} & \cdots & a_{1N} \\ a_{21} & a_{22} & a_{23} & \cdots & a_{2N} \\ \vdots & \vdots & \vdots & \vdots & \vdots \\ a_{N1} & a_{N2} & a_{N3} & \cdots & a_{NN} \end{bmatrix}$$

with interval $[a_{11}, b_{11}]$ as an element in position 1,1. We can start the process with a matrix A_0 such that

$$A_0 = \begin{bmatrix} a_{11} & a_{12} & a_{13} & \cdots & a_{1N} \\ a_{21} & a_{22} & a_{23} & \cdots & a_{2N} \\ \vdots & \vdots & \vdots & \vdots & \vdots \\ a_{N1} & a_{N2} & a_{N3} & \cdots & a_{NN} \end{bmatrix}$$

1. We can compute $\sigma_1(A_0)$, $\sigma_N(A_0)$ the maximal and minimal singular values of A_0. If an inequality $\sigma_1(A_0)\mu^* < \sigma_N(A_0)$ is true then it guaranteed that the given interval matrix is non-practically invertible, i.e. $\mu(\mathbb{A}) > \mu^*$, and the process stop with this inequality as an answer.

2. We can choose a natural number M and real number Δ such that the inequalities

$$\Delta = \frac{b_{11} - a_{11}}{M}, \quad \Delta < \frac{||A_0|| + b_{11} - a_{11}}{4\mu^*}$$

are true.

Denote ($k = 1, 2, ..., M$)

$$A_k = \begin{bmatrix} a_{11} + k\Delta & a_{12} & a_{13} & \cdots & a_{1N} \\ a_{21} & a_{22} & a_{23} & \cdots & a_{2N} \\ \vdots & \vdots & \vdots & \vdots & \vdots \\ a_{N1} & a_{N2} & a_{N3} & \cdots & a_{NN} \end{bmatrix}$$

We can compute $\sigma_1(A_k)$, $\sigma_N(A_k)$ the maximal and minimal singular values of A_k. If an inequality $\sigma_1(A_k)\mu^* < \sigma_N(A_k)$ is true then it guaranteed that the given interval matrix is non-practically invertable, i.e. $\mu(A) > \mu^*$, and the process stop with this inequality as an answer. In another case based on theorem 2 we can conclude that $\mu(\mathbb{A}) < 1.5\mu^*$, i.e. the given interval matrix is practically invertible.

Really, it is clear that $||A_k|| < ||A_0|| + b_{11} - a_{11}$ and $\Delta < ||A_k||/[4\mu^*]$. It is easy to check that $(k = 1, 2, ..., M)$

$$||A_k - A_{k-1}|| = \Delta.$$

Hence in view of the theorem2 we can say that if $\mu(A_k) < \mu^*$ and $(k-1)\Delta \leq$

$\beta \leq (k+1)\Delta$ then $\mu(A(\beta)) < 1.5\mu^*$. Here

$$A(\beta) = \begin{bmatrix} a_{11}+\beta & a_{12} & a_{13} & \cdots & a_{1N} \\ a_{21} & a_{22} & a_{23} & \cdots & a_{2N} \\ \vdots & \vdots & \vdots & \vdots & \vdots \\ a_{N1} & a_{N2} & a_{N3} & \cdots & a_{NN} \end{bmatrix}$$

In some steps of the algorithm we can compute the matrices and the refinement their for the neighboorhood matrices by Zubov's method [4].

References

1. Neumaier, Arnold, Interval methods for System of Equations. Cambridge University Press, 1990.
2. Moore R.E., Interval Analysis. Englewood Cliffs, N.J., Prentice-Hall, 1966.
3. Godunov S.K., Antonov A.G., Kiriluk O.P., Kostin V.I., Guaranteed accuracy for the solving of linear systems in Euclidean spaces. Novosibirsk, Nauka, 1988, 456 p. (in Russian).
4. Zubov S.V., New Algorithm of Matrices Inversion. IMACS/GAMM SCAN-95 International Symposium on Scientific Computing, Computer Arithmetic and Validated Numerics, Wuppertal, September 26-29, Abstracts, 148 p.

Translated by the author

Chapter 7

The doubling algorithms

Haydar Bulgak[1]

Abstract. The most important part of the computations with guaranteed accuracy are the algorithms. In this paper are disscussed the doubling algorithms whose rate of convergence is like the Newton-type algorithms.

1. Zubov's method (iterative refinement of an inverse matrix)

For iterative refinement of the computed approximation of an inverse matrices in [1] was introduced one algorithm. This method allows to compute a matrix $(A+B)^{-1}$ if A^{-1} is known and $||B||$ is sufficiently small ($||\cdot||$ is a spectral norm of the matrices

$$(A+B)^{-1} = A^{-1}A(A+B)^{-1} = A^{-1}[(A+B)A^{-1}]^{-1} = A^{-1}[I+BA^{-1}]^{-1}.$$

It is well known that if $||C|| < 1$ then $[I+C]^{-1}$ exists and has following representation

$$[I+C]^{-1} = \sum_{k=0}^{\infty} C_k.$$

For approximation this series suggested in [1] to use following sums

$$C_m = \sum_{k=0}^{2^m} C_k.$$

We have good formulas for computing these sums ($k = 1, 2, \ldots, m$):

$$X_0 = I, \quad Y_0 = C;$$

$$X_k = X_{k-1} + Y_{k-1}X_{k-1}Y_{k-1}; \quad Y_k = [Y_{k-1}]^2.$$

[1] Selcuk University,Research Center of Applied Mathematics, Konya, Turkey, BULGK@karatay.cc.selcuk.edu.tr

These formulas were suggested by Devison and Man [2] for computation Lyapunov matrix equation. It is clear that

$$||[I+C]^{-1} - C_m|| \leq \sum_{k=2^m+1}^{\infty} ||C||^k = \frac{||C||^{2^m+1}}{1-||C||}.$$

Hence, the process beginning with $C = BA^{-1}$ then it allows us to compute $A^{-1}X_m$ which is an approximation of $(A+B)^{-1}$ and inequality

$$||A^{-1}X_m - (A+B)^{-1}|| \leq \frac{1}{\sigma_1(A)} \frac{(\mu(A)||B||/||A||)^{2^m+1}}{1-\mu(A)||B||/||A||}$$

is true. Here $\mu(A) = ||A||\ ||A^{-1}|| = \dfrac{\sigma_N(A)}{\sigma_1(A)}$ is a condition number of a matrix A, $\sigma_N(A)$, $\sigma_1(A)$ are the maximal and minimal singular values of A. Since if $\dfrac{\mu(A)||B||}{||A||} < 0.5$ then the rate of convergence is given by inequality

$$||A^{-1}X_m - (A+B)^{-1}|| \leq \frac{2}{\sigma_1(A)}(\mu(A)||B||/||A||)^{2^m+1}.$$

This algorithm may be used also for computing the spectral portraits of the matrices. In [3] suggested this method used for checking of a well conditioning of an interval matrix.

2. Solving of the Lyapunov matrix equation

Similar formulas are used for computation a positive definite solution of the Lyapunov matrix equation

$$A^*H(C) + H(C)A + C = 0, C = C^* > 0.$$

If A is an asymptotically stable matrix (see [2]) then the solution in question exists and has a representation

$$H(C) = \int_0^\infty e^{tA^*}Ce^{tA}dt.$$

It is clear that the sequence $(j \geq 0)$

$$H_j(C) = \int_0^{2^j h} e^{tA^*}Ce^{tA}dt$$

for any positive number h is convergent on the class of asymptotically stable A. As a measure of an asymptotically stability of A we can use a number [4],[5]

$$\kappa(A) = 2||A|| \sup_{\dot{x}=Ax,||x(0)||=1} \int_0^\infty ||x(t)||^2 dt.$$

It is important that for matrix exponential e^{tA} we have the estimations ($t > 0$)

$$||e^{tA}|| \leq \sqrt{\kappa(A)} e^{-t \frac{||A||}{\kappa(A)}}.$$

This allows us to conclude that ($j \geq 0$)

$$||H(C) - H_j(C)|| \leq \kappa(A) e^{-2^{j+1} h \frac{||A||}{\kappa(A)}} ||H(C)||$$

we have estimation of the algorithm speed of convergent.

This scheme of computation $H(C)$ in question was introduced by Devison E.P. and Man F.T. [2]. They suggested the following formulas for its computations:

$$B_0 = e^{hA}; \quad H_0 = \int_0^h e^{tA^*} C e^{tA} dt;$$

$$H_k(C) = H_{k-1}(C) + [B_{k-1}]^* H_{k-1}(C) B_{k-1}; \quad B_k = [B_{k-1}]^2.$$

In [5] it was shown that if choose $h = 1/(2||A||)$ then on the set of the asymptotically stable matrices A the formulas are stable to round-off errors and its rate of convergence depend on quality of stability $\kappa(A)$, and on the accuracy of the computer used. The latter is characterised by a small positive number ϵ_1 such that there are no any machine numbers between 1 and $1 + \epsilon_1$.

3. Solving of Lourier-Riccati matrix equation

In this section we describe algorithm with guaranteed accuracy for solving Lourier-Riccati matrix equation

$$A^*\Lambda + \Lambda A + 2||A|| \, [C^*(CFC^*)^{-1}C - \Lambda B(B^*GB)^{-1}B^*\Lambda] = 0,$$

where $F = F^* > 0$ is an $NF \times NF$ symmetric positive-definite matrix, $G = G^* > 0$ is an $NG \times NG$ matrix and A, B, C are the matrices with dimension $N \times N, NF \times N, N \times NC$ respectively. This equation plays an important role in the Automatical Control Theory. It is well-known (see, for example [8]), if a couple (A, B) is a stabilizable and a couple (C, A) is a detectable then there exists Λ a unique symmetric nonnegative definite solution solution of the given matrix equation. The algorithm was suggested and founded in [6]. This algorithm based on the computation the two-point boundary value problem. It includes a series of orthogonal transformations of the matrix made up of the Lourier-Riccati equation coefficients. As a result we obtain the system of equations whose condition is not worse than the condition of the Lourier-Riccati matrix equation. As the measure of condition we use four numerical characteristics suggested by S.K.Godunov [7] which are the norms of the solution of the four matrix equations: the

$$MA^* + AM + 2||A|| \, [B(B^*GB)^{-1}B^* - MC^*(CFC^*)^{-1}CM] = 0,$$

and two Lyapunov matrix equations

$$HK^* + KH + 2||A||\, G^{-1} = 0,$$

$$EL + L^*E + 2||A||\; ||C^*(CFC^*)^{-1}C|| I = 0,$$

where

$$K = A - 2||A||\, B(B^*GB)^{-1}B^*\Lambda,$$

$$L = A - 2||A||\, MC^*(CFC^*)^{-1}C.$$

A complete analysis of the influence of round-off errors and uncertainties in the data has been made [6]. The analysis has shown that the algorithm is as fast as the known algorithm of the sign-function type but is more stable.

A condition number of the given matrix equation was defined by S.K.Godunov [7]. If K, L are the asymptotically stable matrices then we can use the quantities $||\Lambda||, ||M||, ||H||, ||E||$ as a condition number of the given matrix equation.

By the way, from asymptotical stability L and K matrices we have the inequalities

$$||A||\, \frac{\kappa(K^*)}{||G||\; ||K||} \le ||H|| \le ||A||\, \frac{\kappa(K^*)}{||K||}$$

$$||A||\; ||F|| \frac{\kappa(L)}{||L||} \le ||E|| = 2||A||\; ||C^*(CFC^*)^{-1}|| \le ||A||\; ||F^{-1}|| \frac{\kappa(L)}{||L||}.$$

Consequently, alongside the set $||\Lambda||$, $||M||$, $||H||$, $||E||$ for evaluating the conditions of the Lourier-Riccati matrix equation it is possible to use $||A|| \frac{\kappa(K^*)}{||K||}$ and $||A|| \frac{\kappa(L)}{||L||}$ instead $||H||$ and $||E||$ For computating Λ approximations it is convenient to use a two-point boundary value problem [7]: $0 \le \tau \le T$,

$$\frac{d}{dt}\begin{bmatrix} X(t) \\ Y(t) \end{bmatrix} = W \begin{bmatrix} X(t) \\ Y(t) \end{bmatrix}, \mathbb{L}\begin{bmatrix} X(0) \\ Y(0) \end{bmatrix} = I, \mathbb{R}\begin{bmatrix} X(T) \\ Y(T) \end{bmatrix} = 0,$$

where

$$W = \begin{bmatrix} A & 2||A||B(B^*GB)^{-1}B^* \\ 2||A||C^*(CFC^*)^{-1}C & -A^* \end{bmatrix};$$

$$\mathbb{L} = \begin{bmatrix} I_{2N} & 0 \end{bmatrix}; \mathbb{R} = \begin{bmatrix} 0 & I_{2N} \end{bmatrix}.$$

We can represent the following algorithm [6].

1) With the help of the Taylor series at $\tau = 0.5/||W||$ we obtain

$$A_0 = e^{-\tau W},$$

and suppose $B0 = I$.

2) We carry out a successive recurrent determination of the matrix pairs $A_1, B_1; A_2, B_2; \ldots; A_m, B_m; \ldots$ by means of the procedure which is reduced at

every step of the recursion to the choice of the orthogonal transformation P_j, such as

$$P_j = \begin{bmatrix} B_{j-1} & A_{j-1} & 0 \\ A_{j-1} & 0 & B_{j-1} \end{bmatrix} = \begin{bmatrix} \times & \dots & \dots & \dots \\ & \ddots & \dots & \dots \\ 0 & & \times & \dots \end{bmatrix} = \begin{bmatrix} \dots & \dots & \dots \\ 0 & A_j & B_j \end{bmatrix}$$

After l steps of the algorithm we have equations

$$A_{11}^{(l)} X(T) + B_{12}^{(l)} Y(0) + B_{11}^{(l)} = 0;$$

$$A_{21}^{(l)} X(T) + B_{22}^{(l)} Y(0) + B_{21}^{(l)} = 0,$$

where $A_{ij}^{(l)}, B_{i,j}^{(l)}$ - the $N \times N$ matrix elements of the matrices A_l, B_l.

3) Choice of the orthogonal transformation Q, such that

$$Q = \begin{bmatrix} A_{11}^{(l)} & B_{12}^{(l)} & B_{11}^{(l)} \\ A_{21}^{(l)} & B_{22}^{(l)} & B_{21}^{(l)} \end{bmatrix} = \begin{bmatrix} \times & \dots & \dots & \dots \\ & \ddots & \dots & \dots \\ 0 & & \times & \dots \end{bmatrix} = \begin{bmatrix} \dots & \dots & \dots \\ 0 & R_l & F_l \end{bmatrix}$$

where R_l is a upper triangular.

It turns out that $R_l F_l \to \Lambda$ as $\lambda \to \infty$.

Thereat invertibility R_l at rather large l. Furthermore, the velocity of the convergency of the approximations to Λ is estimated beginning from some l by the inequalities of the form:

$$||R_l^{-1} F_l - \Lambda|| \leq ||2\Lambda|| \, [1 + ||M|| \, ||\Lambda||] \kappa(K^*) e^{-2^{j+1} T \frac{||K||}{\kappa(K^*)}}.$$

The proposed algorithm is iterative. On all its next steps the interval of the correspond boundary value problem is doubling. The rate of convergence depends on the condition of the Lourier Riccati matrix equation.

References

1. Zubov S.V., New Algorithm of Matrices Inversion. IMACS/GAMM SCAN 95 International Symposium on Scientific Computing, Computer Arithmetic and Validated Numerics, Wuppertal, September 26-29, Abstracts, 148 p.
2. Davison E.J., Man F.T., The numerical solution of $A^*Q + QA = -C$. IEEE Trans. Aut. Control AC-13, 1968, pp. 448–449.
3. Ayshe BULGAK, Checking of a Well-conditioning of an Interval Matrix. Siberian J. of Dif. Equations (To appear).
4. Bulgakov A.Ya., An effectively calculable parameter for the stability quality of systems of linear differential equations with constant coefficients. in Siberia Math. J., N. 21, 1980.

5. Bulgakov A.Ya., Godunov S.K., Calculation of positive definite solutions of Lyapunov's equation. Numerical methods of linear algebra, Novosibirsk, Nauka, 1985, pp .17–38. (in Russian)

6. Bulgakov A.Ya., The quaranteed accuracy algorithm for solving Lourier-Riccati matrix equation. Teoreticheskiye i Vichislitelniye Problemi Matematicheskoi Fiziki (Теоретические и вычислительные проблемы математической физики)б Institute of Mathematics of Siberian Division Russian Academy of Science, Novosibirsk, 1994, pp. 63–104 (in Russian).

7. Godunov S.K., Norms of the solutions of Lourier-Riccati matrix equations as criteria of the quality of stabilizability and detectability. J. Siberian Adwances in Mathematics, New-York, V. 2, N. 3, 1992, pp. 135–157.

8. Wonham W.M., Linear multivariable control: a geometric approach. Springer-Verlag, New-York, 2-nd edition, 1979.

Translated by the author

Chapter 8

Dynamics of "society-police" system

L.V. Nedorezov[1] and T.L. Schedrina[2]

Abstract. In this paper we consider the mathematical model of dynamics of the society's number. A society is divided on two various categories of citizens: the serving in the police and all other ones. It is assumed the number of the policemen and their level of the equipment exert significant influence of the parameters of change of the society's number. The dynamic regimes of the model and the conditions of appearance of the critical situations are investigated when the level of social tension in the society reaches the extreme significances.

Introduction

The various structures such as the police, the church, the army exert direct influence on dynamics of the society's number, on the level of social tension. This influence tells on the values of all the main parameters such as the birth rate, the death rate (as the natural death rate defined only by the biological characteristics of the individuals as the additional death rate depending on the conditions of the environment, the rate of spreading of various diseases, the level criminality etc.), creation material wealth and so on. Therefore it is interesting to analyse the influence of those or other structures on dynamics of society in "pure form" when only one such a structure is selected in the model.

In the present work it is considered one of the managing structures — the police. Constructing the model we shall proceed from assumptions that increase of the number of the policemen results in decrease of the level of social tension (as the result of increase of a social safety, the isolation of the criminal elements, the preventive measures etc.), but at that the maintenance costs of the policemen increase (salary, equipment) and volume of the created material wealth decreases (as far as the employees of the police do not produce the material wealth). Except for these assumptions we shall do other ones permitting to construct the model "in first approximation" in the form of the system of four nonlinear ordinary differential equations.

The main purpose of the researches consists in determination the typical change of the society's number and the level of social tension at various values

[1]Institute of Medical and Biological Cybernetics of Russian Academy of Sciences, Siberian Branch, Novosibirsk, Russia

[2]Novosibirsk State University, Novosibirsk, Russia

of the model parameters.

1. The description of model

Assume that the description of the number of society is based on the Verhulst's model [1] modified for case when society is divided on two groups of the individuals: the employees in the police (the number of which we shall designate through $y(t)$ at the moment of time t) and all other (the number of which we shall designate through $x(t)$):

$$\begin{cases} \dfrac{dx}{dt} = \alpha_1(x+y) - \alpha_2 x - \beta x(x+y) - \alpha_3 x + \alpha_4 y, \\ \\ \dfrac{dy}{dt} = \alpha_3 x - \alpha_4 y - \alpha_2' y - \beta y(x+y). \end{cases} \tag{1}$$

In the systems (1) the expression $\alpha_1(x+y)$ is equal to the rate of increase of the number of population not serving in the police; α_2, α_2' are the intensions of the natural death rate of the corresponding groups of society which is different in general case. Let us shall suppose further that $\alpha_2' > \alpha_2$. Assume that the quantities α_1, α_2 and α_2' are the positive constants not depending on the influence on the considered managing structure. This supposition is true only in the first approximation: in general case the influence of the given structure results in the isolation of part of the individuals that doubtless affects on the birth rate (the rate decreases) and on the death rate of the individuals (the death rate evidently increases).

The expression $\psi = \beta x(x+y)$ in (1) corresponds to the "additional" death rate connected with influence of the ecological factors (overpopulation, pollution of environment etc.), the diseases, the level of the criminality and so on. Let us assume that in considered case the function β depends on two quantity — the part of the policemen in society $\xi = \dfrac{y}{x+y}$ and "equipment" of the police B (armament, transport, buildings etc.) and satisfies the following conditions:

$$\begin{array}{l} \beta = \beta(B,\xi), \ \dfrac{\partial \beta}{\partial \xi} < 0, \ \dfrac{\partial \beta}{\partial B} < 0, \\ \exists \beta^*, \beta_0 \equiv \text{const} > 0 \ : \ \forall B, \xi \quad \beta_0 \le \beta(B,\xi) \le \beta^* < \infty \\ \beta^* = \beta(B,0). \end{array} \tag{2}$$

The conditions (2) mean that the value of the quantity β decreases with increase of the number of the policemen and with increase their equipment and, as consequence, decrease the additional death rate. Besides there are boundaries of change of the quantity β. In particular the maximum value β^* is reached in that case when the number of the police is equal to zero in society.

Note that it is possible to interpret expression $\beta x(x+y)$ as the level of social tension. In the most simple case the function β can be presented as

$$\beta = \frac{\beta_1}{1+\beta_2 B\xi} + \beta_0. \tag{3}$$

The coefficients α_3 and α_4 in the system (1) describe the migrated flows (entry into the police and dismissal with service). Presume that the entry on service is defined by the level of social tension ψ and this flow increases with increase of the quantity ψ

$$\begin{aligned} &\alpha_3 = \alpha_3(\psi) \geq 0, \quad \frac{d\alpha_3}{d\psi} > 0, \quad \alpha_3(0) = 0, \\ &\exists \alpha_3^* \equiv \text{const} < \infty : \quad \forall \psi \ , \alpha_3(\psi) \leq \alpha_3^*. \end{aligned} \tag{4}$$

In the most simple case it is possible to choose α_3 in the form of the linear-fractional function

$$\alpha_3 = \frac{\alpha_3^1 \psi}{\alpha_3^2 \psi + 1}, \quad \alpha_3^1, \alpha_3^2 \equiv \text{const} > 0. \tag{5}$$

In the frameworks of the model let us consider that the value of the coefficient α_4 (accordingly the rate of dismissal from service) depends on two main the quantities — the average equipment for one separate policeman $\zeta = \dfrac{B}{y}$ and a tension of spending of money resources on salary γ_2 (i.e., $\gamma_2 y \Delta t$ is a quantity of money separated on salary to the policemen for time Δt). Naturally to assume that increase of salary and the average equipment results in decrease the corresponding migrated flow

$$\begin{aligned} &\alpha_4 = \alpha_4(\gamma_2, \zeta), \quad \frac{\partial \alpha_4}{\partial \gamma_2} < 0, \quad \frac{\partial \alpha_4}{\partial \zeta} < 0, \\ &\exists \alpha_4^* \equiv \text{const} < \infty : \quad \forall \gamma_2, \zeta \ \ \alpha_4(\gamma_2, \zeta) \leq \alpha_4^*. \end{aligned} \tag{6}$$

Further at research of the model we use the following function satisfying to the conditions (6)

$$\alpha_4 = \frac{\alpha_4^1}{1 + \alpha_4^2 \gamma_2 + \alpha_4^3 \zeta}. \tag{7}$$

The payment of salary to the employees of the police depends on the state of the money fund, volume of which at the moment of time t we designate as $W(t)$. At satisfactory state of this fund when its volume exceeds some quantity $W_2 \equiv \text{const} > 0$ let us suppose that the salary is constant (it is established in the legislative order) and monotonically decreases otherwise (in particular as the result of the delayed payment of salary, the conversion on the partial working day, decrease of every possible extra charges to salary etc.). At analysis of the model we consider the following function

$$\gamma_2 = \begin{cases} \gamma_2^0 \sqrt{W_2} & \text{at } W \geq W_2 \\ \gamma_2^0 \sqrt{W} & \text{at } W < W_2, \quad \gamma_2^0 \equiv \text{const} > 0. \end{cases} \tag{8}$$

The dynamics of volume $W(t)$ money fund of the police is defined by three main following processes: by the rate of supplement of money resources in the fund $\gamma_1 x$ which is direct proportionally to the number of the people not working in the police, to the level of social tension and in inverse proportion to volume of money fund; by the rate of charge of money resources on salary $\gamma_2 y$ and by the rate of charges on the equipment of the police γ_3

$$\frac{dW}{dt} = \gamma_1 x - \gamma_2 y - \gamma_3. \tag{9}$$

The function γ_1 in (9) pursuant to its meaning load satisfies the following conditions

$$\begin{array}{c} \gamma_1 = \gamma_1(\psi, W), \ \dfrac{\partial \gamma_1}{\partial \psi} \geq 0, \ \dfrac{\partial \gamma_1}{\partial W} \leq 0, \\ \exists \gamma_1^* \equiv \text{const} < \infty \ : \ \forall \psi, W \ \ \gamma_1 \leq \gamma_1^* \end{array} \tag{10}$$

and in the most simple case it can be presented as the linear-fractional function

$$\gamma_1 = \frac{\gamma_1^2 \psi}{1 + \gamma_1^3 W + \gamma_1^4 \psi}, \quad \gamma_1^2, \gamma_1^3, \gamma_1^4 \equiv \text{const} > 0. \tag{11}$$

As to the quantity γ_3 (9) flow of money resources on the equipment of the police let us shall assume an existence of some critical value $W_3 \equiv \text{const} > 0$ such that if the volume of the fund $W(t)$ exceeds W_3 then the rate γ_3 is constant γ_3^0 else it is equal to zero, i.e.

$$\gamma_3 = \gamma_3^0 \theta(W - W3), \tag{12}$$

where θ — the Heaviside theta-function .

Designate through $B(t)$ the equipment of the police (in the money form) at the moment of time t. The rate of change of this quantity is formed from the following processes: the receipt of money resources from fund $W(t)$ and decrease connected with depreciation of the equipment, the buildings, the uniforms, the charge of the fuel, the ammunitions etc. Assume that decrease of the quantity $B(t)$ is direct proportionally to the quantity B and thus its dynamics is defined equation

$$\frac{dB}{dt} = \gamma_3 - \pi B, \tag{13}$$

where the intensity of decrease π satisfies conditions

$$\pi = \pi(y), \quad \pi(0) > 0, \quad \frac{d\pi}{dy} > 0. \tag{14}$$

The conditions (14) mean that the tension of decrease of the police's equipment depends on the number of the policemen and increases with increase their number.

In the most simple case the function $\pi(y)$ is linear

$$\pi = \pi_1 y + \pi_2, \quad \pi_1, \pi_2 \equiv \text{const} > 0. \tag{15}$$

Uniting the equations (1), (9), (13) with the conditions (2)–(8), (10)–(12) and (14), (15) we receive the required model of the system "society-police" in the most simple form. The model is nonlinear even in the given elementary form and besides it is defined not in all points of the non-negative part of the phase space R_+^4.

2. The analysis of the model

The model of the system "society-police" described in previous section has the following properties:

1. There is a stable invariant compactum Δ in R^4_+:
$\Delta = [0, x^*] \times [0, y^*] \times [0, B^*] \times [0, W^*]$,
where $x^* = \dfrac{\alpha_1 + \sqrt{\alpha_1^2 + 4\beta_0 y^*(\alpha_1 + \alpha_4^*)}}{2\beta_0}$; $y^* = \dfrac{\alpha_3^*}{\beta_0}$; $B^* = \dfrac{\gamma_3^0}{\pi(0)}$;
W^* — the solution of the equation $\gamma_1(\psi^*, W)x^* - \gamma_3(W) = 0$ (it exists and is unique).

Thus the solutions of the considered system of differential equations are non-negative confined and continued in time.

2. The area $\omega = \{(0, 0, W, 0) \mid 0 \leq W \leq W_3\}$ is set of the points of equilibrium of the system. Note that for $\alpha_1 < \alpha_2$ the trajectories of the system tend to ω in asymptotics. The "saddle" movement of trajectories is observed in environs of the beginning of coordinates in case $\alpha_1 > \alpha_2$.

3. At the numerical research of the system two various dynamic regimes were revealed only: the asymptotic degeneration (fig.1) and the stabilization of all the variable of the model on the unique nontrivial level (fig.2).

In case of degeneration (fig.1) for $B(0) = 0$ at first we observe increase of the police's equipment with subsequent sharp decrease of this quantity $B(t)$, $B(t) \to 0$ $t \to \infty$. Accordingly, it is cause for the primary decrease of the level of tension in society.

In second case (fig.2) (when the birth rate is enough high) two qualitatively various regimes were observed: the tension in society tends to the initial value in asymptotics (when, for example, $B(t) \to 0$ and the situation appears when there are enough many the policemen without any of equipment) or when the tension is considerably decrease.

3. Conclusion

The analysis of dynamics of the system "police-society" has shown that the level of social tension increases in the degenerated society. Increase of tension is possible in the stable society, but at that the unnatural situation emerges when the number of the policemen exceeds the summary number of all other stratums of society and at that the level of the equipment of the employees of the police tends to zero.

Also in the frameworks of this model were found the "normal" regimes of functioning of the system when we have a situation with the natural proportions of the number of the considered social stratums and with enough low level of social tension.

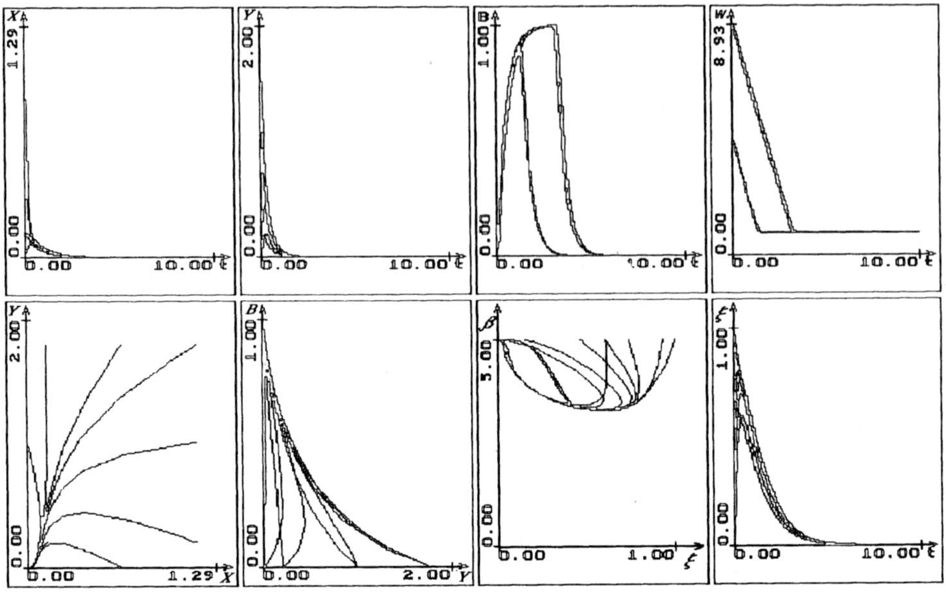

Figure 1: The regime of degeneration.

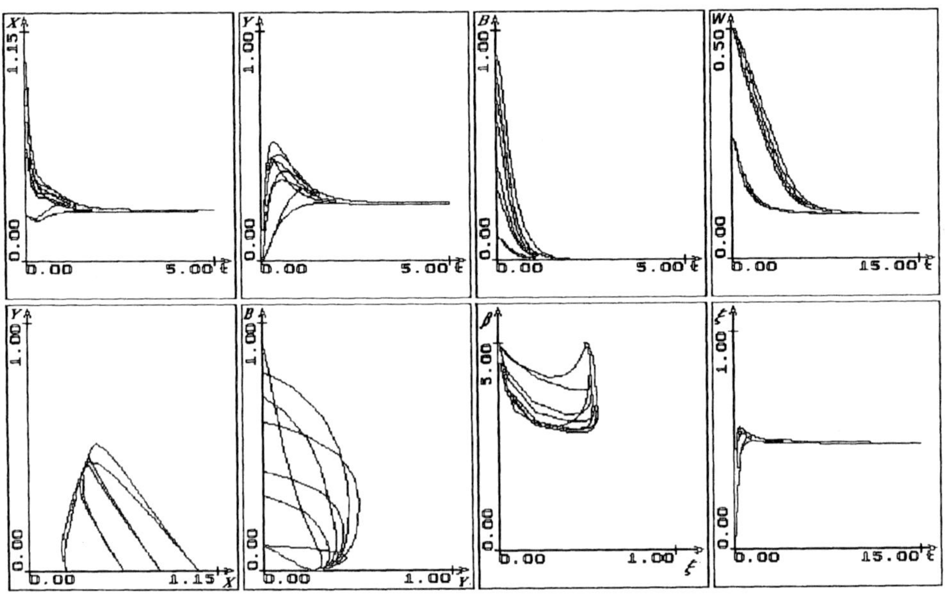

Figure 2: The regime of stabilization on nontrivial level.

References

1. Verhulst P.F., Notice sur la loi que la population suit dans son accroissement. Corresp.Math. Et Phys., 10, 1838, pp. 113–121.

Translated by the authors

Chapter 9

A second order scheme for nonlinear singularly perturbed two-point boundary value problem

A.I.Zadorin[1]

Abstract. An nonlinear second order ordinary singularly perturbed equation is consided.We suppose that the solution has boundary layers near the endpoints of the interval. We use the linear interpolation for the source in the equation for the every mesh interval. The concordance of the solutions in the neighbouring mesh intervals leeds to the finite-difference scheme. We prove that under some conditions this scheme has the second order of the accuracy uniformly in a small parameter.

We shall consider the following singularly perturbed boundary value problem:

$$Lu = \varepsilon u'' - f(x,u) = 0, \quad u(0) = A, \quad u(1) = B \tag{1}$$

with the hypotheses, which will be assumed throughout the paper:

$$\varepsilon > 0, \quad f \in C^2(I,R), \quad I = [0,1], \quad f'_u \geq \beta > 0. \tag{2}$$

The solution $u(x)$ has two boundary layers near the endpoints of the interval I.

In the linear case of problem (1) the question of construction of a difference scheme with the property of uniform convergence in a small parameter ε at first was considered by Bahvalov N.S [1]. In that paper was proved the uniform convergevce of the central difference scheme on the mesh with the points concentrated in the boundary layers. This approach, based on the concentration of mesh points in the regions where the solution changes rapidly, was developed by Liseikin V.D., Petrenko B.E. [2]. In [3] the numerical soluton of nonlinear problem (1) was found by using Richardson extrapolation on a special nonuniform discretizaton mesh.In [4] after some transformation the problem (1) was solved by a standard difference scheme on the equidistant mesh in the combination with the solution of the corresponding reduced problem.

Another approach for construction of the schemes with the property of uniform convergence in a small parameter ε is to take into account boundary-layer behavior of the solution in the coefficients of the difference scheme.

[1]Institute of information technologies and applied mathematicians of Subirian department of Russian academy of science, Omsk

An exact difference scheme was founded by Samarsky A.A. [5]. Familiar approach was presented by Gaevoy V.P. [6]. In the linear case the special difference scheme was constructed in [7], [8] and in some other papes [9].

In this paper for nonlinear problem (1) we'll construct the scheme of the second order accuracy uniformly in ε . Our method corresponds to the second from the mentioned approaches. Throughout the paper C and C_i denote positive constants independing of ε and mesh steps. We use the maximum norm $||V|| = \max|V_i|$ for any vector $V = (V_1, V_2, ..., V_N)$ and $||V|| = \max|V(x)|$, $x \in I$ for any function $V(x)$.

We introduce the mesh Ω:

$$\Omega = \{x_i : x_i = x_{i-1} + h_i, x_0 = 0, x_N = 1\}, \quad \Delta_i = [x_i, x_{i+1}).$$

To construct the difference scheme for the problem (1) we consider the problem:

$$\widetilde{L}V = \varepsilon V'' - \widetilde{f}(x, V) = 0, \quad V(0) = A, \quad V(1) = B, \tag{3}$$

where for $x \in \Delta_n$

$$\widetilde{f}(x, V) = f_n + f'_{xn}(x - x_n) + f'_{Vn}(V - V_n)$$

with

$$f_n = f(x_n, V_n), \quad f'_{xn} = f'_x(x_n, V_n), \quad f'_{Vn} = f'_V(x_n, V_n), \quad V_n = V(x_n).$$

For every interval Δ_n the solution of the problem (3) has a form:

$$V_n(x) = V_n^0(x) + \gamma_1^{(n)} \exp[\alpha_n(x - x_n)] + \gamma_2^{(n)} \exp[-\alpha_n(x - x_n)], \tag{4}$$

where V_n^0 is some solution of the equation (3):

$$V_n^0(x) = -\beta_n^{-2}(b_n x + d_n) + \beta_n^{-2}(b_n x_n + d_n)\cosh[\alpha_n(x - x_n)] +$$
$$+ b_n \beta_n^{-2} \alpha_n^{-1} \sinh[\alpha_n(x - x_n)],$$

where

$$\alpha_n = \beta_n \varepsilon^{-1/2}, \quad \beta_n = (f'_{Vn})^{1/2}, \quad b_n = f'_{xn}, \quad d_n = f_n - b_n x_n - f'_{Vn} V_n.$$

We select $\gamma_1^{(n)}, \gamma_2^{(n)}$ to satisfy conditions of continuity for $V(x)$:

$$V_n(x_n) = V_n^h, \quad V_n(x_{n+1}) = V_{n+1}^h.$$

The condition $V \in C^1[0, 1]$ is valid, if

$$V'_{n-1}(x_n) = V'_n(x_n), \; n = 1, 2, ..., N - 1.$$

Using (4), we obtain the system of finite-difference equations:

$$\frac{\beta_n}{\sinh(\alpha_n h_{n+1})} V_{n+1}^h - \left[\frac{\beta_n}{\sinh(\alpha_n h_{n+1})} + \beta_{n-1}\coth(\alpha_{n-1} h_n)\right] V_n^h +$$

$$+\beta_{n-1}\coth(\alpha_{n-1}h_n)V_{n-1}^n = F_n,$$

$$V_0^h = A, \quad V_N^h = B, \quad n = 1, 2, ..., N-1, \tag{5}$$

where

$$F_n = \frac{f_{n-1}}{\beta_{n-1}}\tanh\frac{\alpha_{n-1}h_n}{2} + \frac{f_n}{\beta_n}\tanh\frac{\alpha_n h_{n+1}}{2} +$$

$$+\frac{b_n h_{n+1}}{\beta_n}\left[\frac{1}{\alpha_n h_{n+1}} - \frac{1}{\sinh(\alpha_n h_{n+1})}\right] +$$

$$+\frac{b_{n-1}h_n}{\beta_{n-1}}\frac{\alpha_{n-1}h_n\coth(\alpha_{n-1}h_n) - 1}{\alpha_{n-1}h_n}.$$

The difference scheme (5) is a system of nonlinear algebraic equations. The solution of this system is bounded, if some conditions on the function f are assumed.

Lemma 1. *Suppose that*

$$f = f(u), \quad f''(u)u \geq 0. \tag{6}$$

Then

$$||V^h|| \leq \max\{|A|, \ |B|, \ |f(0)|/\beta\}. \tag{7}$$

Proof. Rewrite scheme (5) in a form:

$$L_n^h V^h = \frac{\beta_n}{\sinh(\alpha_n h_{n+1})}V_{n+1}^h -$$

$$-\left[\frac{\beta_n}{\sinh(\alpha_n h_{n+1})} + \beta_{n-1}\coth(\alpha_{n-1}h_n) + \frac{f'(\theta_n)}{\beta_n}\tanh\frac{\alpha_n h_{n+1}}{2}\right]V_n^h +$$

$$+\left[\beta_{n-1}\coth(\alpha_{n-1}h_n) - \frac{f'(\kappa_n)}{\beta_{n-1}}\tanh\frac{\alpha_{n-1}h_n}{2}\right]V_{n-1}^h =$$

$$= \frac{f(0)}{\beta_{n-1}}\tanh\frac{\alpha_{n-1}h_n}{2} + \frac{f(0)}{\beta_n}\tanh\frac{\alpha_n h_{n+1}}{2},$$

$$V_0^h = A, \quad V_N^h = B, \quad n = 1, 2, ..., N-1,$$

where $\kappa_n \in (0, V_{n-1}^h)$. Using (6) we can prove that $\beta_{n-1}^2 \geq f'(\kappa_n)$. It follows that maximum principle for operator L^h is valid (see [5], [10]) if

$$L_n^h \Psi^h \leq 0, \quad \Psi_0^h \geq 0, \quad \Psi_N^h \geq 0, \quad n = 1, 2, ..., N-1\ , \tag{8}$$

then $\Psi_n^h \geq 0$, $n = 0, 1, ..., N$. We define $\Psi^h = \max\{|A|, \ |B|, |f(0)|/\beta\} \pm V^h$. For given Ψ^h inequalities (8) have a place, it follows $\Psi^h \geq 0$. Lemma is proved.

Lemma 2. *Suppose that $||V^h|| \leq C$, where V^h is the solution of the scheme* (5)*. Then*

$$||V|| \leq C_1, \quad |V'(x)| \leq C_2\left[1 + \varepsilon^{-1/2}\left(\exp(-\kappa x) + \exp(\kappa(x-1))\right)\right], \tag{9}$$

where $\kappa = (\beta/\varepsilon)^{1/2}$, V *is the solution of the problem* (3).

Proof. Write the problem (3) in a form

$$\varepsilon V'' - a(x)V = g(x), \quad V(0) = A, \quad V(1) = B, \tag{10}$$

where for $x \in \Delta_n$ $a(x) = \beta_n^2$, $g(x) = f_n + b_n(x - x_n) - \beta_n^2 V_n^h$. Taking into account that $||V^h|| \le C$ we get $|g(x)| \le C_3$. Using inequality $||V|| \le |A| + |B| + \beta^{-1}||g||$, we get $||V|| \le C_1$. Using the condition $|g(x)| \le C_3$ and famillar results [7], we get (9). Lemma is proved.

Let Ω be a nonuniform mesh of the interval I. We want a relatively high density of points in a boundary layers. Let M be a number of mesh points in the boundary layer. We take for $n \le M$

$$x_n = -\kappa^{-1} \ln[1 - (1 - \kappa^{-1})M^{-1}n]. \tag{11}$$

For $n \ge N - M$

$$x_n = 1 + \kappa^{-1} \ln[1 - (1 - \kappa^{-1})M^{-1}(N - n)]. \tag{12}$$

We suppose, that the mesh Ω is uniform out of boundary layers, Q is a number of uniform steps, $Q = N - 2M$.

Theorem 1. *Let* Ω *satisfies to conditions* (11)–(12), V^h *is a bounded solution of the scheme* (5), $[u]_\Omega$ *is the solution of the problem* (1) *on mesh points. Then*

$$||V^h - [u]_\Omega|| \le C[Q^{-2} + M^{-2}]. \tag{13}$$

Proof. Let $z = u - V$. Then

$$\varepsilon z'' - [f(x,u) - f(x,V)](u - V)^{-1} z = f(x,V) - \widetilde{f}(x,V).$$

For $x \in \Delta_n$ $|f(x,V) - \widetilde{f}(x,V)| \le C[h_{n+1}^2 + |V(x) - V(x_n)|^2]$. Using the estimate (9), we get

$$|V(x) - V(x_n)| \le \int_{x_n}^{x} |V'(s)| ds \le C_1 h_{n+1} +$$

$$+ C_2\{\exp[-\kappa x_n] - \exp[-\kappa x_{n+1}] + \exp[\kappa(x_{n+1} - 1)] - \exp[\kappa(x_n - 1)]\}.$$

For $n < M$ we use (11) and get :

$$|V(x) - V(x_n)| \le C_3[h_{n+1} + M^{-1}].$$

$h_{n+1} < M^{-1}$ for $n < M$. It follows

$$|V(x) - V(x_n)| \le CM^{-1}.$$

The case $n > N - M$ may be considered in the same manner.

For $M \le n \le N - M$ we get $|V(x) - V(x_n)| \le CQ^{-1}$. So, we get

$$|f(x,V) - \widetilde{f}(x,V)| \le C[M^{-2} + Q^{-2}].$$

Taking into account, that $[V]_\Omega = V^h$ and $||z|| \le \beta^{-1}||f(x,V) - \widetilde{f}(x,V)||$, we get (13). Theorem is proved.

Consider the case of a linear problem :

$$Lu = \varepsilon u'' - c(x)u = f(x), \ u(0) = A, \ u(1) = B, \tag{14}$$

where $c(x) \ge \beta > 0$, $\varepsilon > 0$, $f, c \in C^2[0,1]$.

We rewrite scheme (5) for the problem (14):

$$\begin{aligned} L_n^h V^h = & \frac{\beta_n}{\sinh z_{n+1}} V_{n+1}^h - \left[\frac{\beta_n}{\sinh z_{n+1}} + \beta_{n-1}\coth z_n + \right. \\ & \left. + \frac{c_n}{\beta_n}\tanh\frac{z_{n+1}}{2} + \frac{c'_n h_{n+1}}{\beta_n}\left(\frac{1}{z_{n+1}} - \frac{1}{\sinh z_{n+1}}\right)\right] V_n^h + \\ & + \left[\beta_{n-1}\coth z_n - \frac{c_{n-1}}{\beta_{n-1}}\tanh\frac{z_n}{2} - \frac{c'_{n-1}h_n}{\beta_{n-1}}\frac{z_n\coth z_n - 1}{z_n}\right] V_{n-1}^h = \\ = & \frac{f_{n-1}}{\beta_{n-1}}\tanh\frac{z_n}{2} + \frac{f_n}{\beta_n}\tanh\frac{z_{n+1}}{2} + + \frac{f'(x_n)h_{n+1}}{\beta_n}\left[\frac{1}{z_{n+1}} - \frac{1}{\sinh z_{n+1}}\right] + \\ & + \frac{f'_{n-1}h_n}{\beta_{n-1}}\frac{z_n\coth z_n - 1}{z_n}, \end{aligned} \tag{15}$$

where

$$c_n = c(x_n), \quad c'_n = c'(x_n), \quad \beta_n^2 = c_n, \quad \alpha_n^2 = \beta_n^2/\varepsilon;$$
$$z_n = \alpha_{n-1}h_n, \quad f_n = f(x_n), \quad f'_n = f'(x_n).$$

Theorem 2. *Let for any step h_n*

$$h_n \le h, \ h^2||C''|| \le \beta. \tag{16}$$

Then the next estimate for the error of the scheme (15) is valid:

$$||V^h - [u]_\Omega|| \le Ch^2.$$

Proof. For the problem (14) the equation (3) has a form

$$\varepsilon V'' - \widetilde{c}(x)V = \widetilde{f}(x), \quad V(0) = A, \quad V(1) = B, \tag{17}$$

where for $x \in \Delta_n$

$$\widetilde{c}(x) = c_n + c'_n(x - x_n), \quad \widetilde{f}(x) = f_n + f'_n(x - x_n).$$

Prove estimate $||V|| \le C$. Taking into account (16), we have

$$\widetilde{c}(x) = c(x) - c''(\theta)(x - x_n)^2/2 \ge \beta/2.$$

It imply $|V(x)| \leq 2\beta^{-1}\|\widetilde{f}\| + \max\{|A|,|B|\}$. So, $\|V(x)\| \leq C$. Now we estimate $z = u - V$. We have

$$Lz = f(x) - \widetilde{f}(x) + [c(x) - \widetilde{c}(x)]V.$$

Hence, $|Lz| \leq Ch^2$. Taking into account, that $\|z\| \leq \beta^{-1}\|Lz\|$, we get $\|u - V\| \leq Ch^2$. Using that $[V]_\Omega$ is an exact solution of scheme (15), we complete the proof of the Theorem 2.

Now we'll consider the results of the numerical experiments.

Write a boundary value problem in a case of linear equation [4]

$$\varepsilon^2 u'' - u = \cos^2 \pi x + 2(\pi\varepsilon)^2 \cos(2\pi x), \quad u(0) = 0, \quad u(1) = 0. \tag{18}$$

The solution of (18) has a form

$$u(x) = \frac{\exp(-x/\varepsilon) + \exp(-(1-x)/\varepsilon)}{1 + \exp(-1/\varepsilon)} - \cos^2(\pi x).$$

According to the Theorem 2 the scheme (15) has a property of the second order accuracy for any mesh. Let Ω be uniform mesh with the step h. We present a norm of the error $[u]_\Omega - V^h$ depending on ε and h in the table 1. The table 1 confirms the Theorem 2. In the Table 2 is presented the norm of the error for the usiall second order difference scheme. Numerical results confirm the advantage of the scheme (15).

Now we'll consider the case of nonlinear boundary-value problem

$$\varepsilon u'' - u^3 - u = f(x), \ u(0) = 1, \ u(1) = 1, \tag{19}$$

where function $f(x)$ corresponds to solution

$$u(x) = \frac{\exp(-\varepsilon^{1/2}x) + \exp(\varepsilon^{-1/2}(x-1))}{1 + \exp(-\varepsilon^{-1/2})} + \sin(\pi x).$$

The difference scheme (5) is a system of nonlinear algebraic equations. We use modified Pikar method to find its solution [11]

$$\frac{\beta_n^{(k)}}{\sinh(\alpha_n^{(k)} h_{n+1})} V_{n+1}^{(k+1)} - \left[\frac{\beta_n^{(k)}}{\sinh(\alpha_n^{(k)} h_{n+1})} + \beta_{n-1}^{(k)} \coth(\alpha_{n-1}^{(k)} h_n)\right] V_n^{(k+1)} +$$

$$+\beta_{n-1}^{(k)} \coth(\alpha_{n-1}^{(k)} h_n) V_{n-1}^{(k+1)} - G V_n^{(k+1)} = F_n^{(k)} - G V_n^{(k)},$$

$$V_0^{(k+1)} = A, \quad V_N^{(k+1)} = B, \quad n = 1, 2, \dots. N-1, \tag{20}$$

As initial iteration for solution we chose $V_n^{(0)} = 1$ for any n. The numerical experiments indicated the convergence of iterative method for $G > 1$. For calculations we chose $G = 5$. We break iterative process if maximal error between solutions of neighbouring iterations not more 10^{-5}. On every iterative step we used Gaussian elimination method [5].

Table 3 contains a norm of the error for scheme (5) in the case of problem (19) depending on ε and M, where M is number of mesh steps in each boundary layer. A number of uniform steps out of baundary layers is the same as in layers ($Q = M$). In the layers mesh points were chosen according to (11), (12). The results given in Table 3 support estimate (13).

Table 4 contains a norm of the error for scheme (5) in the case of the uniform mesh. This calculations indicate that in the case of nonlinear problem and scheme (5) there is necessity to concentrate mesh points in the boundary layers.

Table 1.

ε	h			
	0.1	0.05	0.01	0.005
1.0	5.9E-2	1.5E-2	6.2E-4	1.5E-4
0.1	3.2E-2	8.2E-3	3.3E-4	8.3E-5
0.01	4.0E-2	9.2E-3	3.3E-4	8.2E-5
0.005	4.3E-2	1.0E-2	3.3E-4	8.3E-5

Table 2.

ε	h			
	0.1	0.05	0.01	0.005
1.0	3.1E-2	7.6E-3	3.0E-4	7.5E-5
0.1	1.2E-2	3.2E-3	1.3E-4	3.3E-5
0.01	9.7E-3	3.0E-2	1.4E-2	3.7E-3
0.005	2.5E-3	9.8E-3	3.6E-2	1.4E-2

Table 3.

ε	M			
	3	10	30	100
0.1	0.16E-1	0.17E-2	0.36E-2	0.60E-3
0.1E-1	0.49E-1	0.42E-2	0.29E-2	0.48E-3
0.1E-2	0.18	0.12E-1	0.33E-2	0.56E-3
0.1E-3	0.29	0.19E-1	0.35E-2	0.60E-3
0.1E-4	0.32	0.23E-1	0.36E-2	0.61E-3

Table 4.

ε	h		
	0.1	0.02	0.01
0.1	0.11E-1	0.33E-3	0.12E-3
0.01	0.71E-1	0.13E-2	0.24E-3
0.001	0.91	0.24E-1	0.38E-2
0.0001	3.3	0.38	0.81E-1

References

1. Bahvalov N.S. About the optimization of methods for solution of boundary value problems with boundary layer. Jornal of comput. math. and math. phys., V. 9, N. 4, 1969, pp. 841–859.
2. Liseikin V.D., Petrenko V.E. The Adaptive invariant method of the numerical solution of problems with boundary and interior layers. Novosibirsk , 1989.
3. Vulanovic R., Herceg D., Petrovic N. On the extrapolation for a singularly perturbed boundary value problem.Computing , V. 36, N. 1–2, 1986, pp. 69–79.
4. Herceg D. On numerical solution of singularly perturbed boundary value problem. Review of Research faculty of science mathematics series, Novi Sad, V. 23, N 1, 1993, pp. 371–381.
5. Samarsky A.A. Theory of difference schemes. Moskow, 1977. itemGaevoy V.P. About one method of difference scheme construction for two-point boundary value problems. Computing systems, Novosibirsk, N. 75, 1978, pp. 96–110.
6. Doolan E.P., Miller J.J.H., Schilders W.H.A. Uniform numerical methods for problems with initial and boundary layers.Dublin, Boole Press, 1980.
7. Zadorin A.I. The difference scheme for self-adjoint singularly perturbed third baundary value problem. Modelling in mechanics, Novosibirsk, V. 3, N. 1, 1989, pp. 77–82.
8. Kadalbajoo M.K., Reddy Y.N. Asymptotic and numerical analysis of singular perturbation problems: a survey. Appl. Math. and Comput., V. 30, N. 3, 1989, pp. 223–259.
9. Zadorin A.I., Ignat'ev V.N. Numerical solution of a singularly perturbed second order quasilinear equation. Jornal of comput. math. and math. phys., V. 31, N. 1, 1991, pp. 157–161.
10. Ortega J.M., Rheinboldt W.C. Iterative solution of nonlinear equations in several variables. New York and London, Academic Press, 1970.

Translated by the author

Index

N

O

P

R

S

T